I0467864

Assessment of Managed Aquifer Recharge from Sand Hollow Reservoir, Washington County, Utah, Updated to Conditions in 2010

By Victor M. Heilweil and Thomas M. Marston

Prepared in cooperation with the Washington County Water Conservancy District

Scientific Investigations Report 2011–5142

U.S. Department of the Interior
U.S. Geological Survey

U.S. Department of the Interior
KEN SALAZAR, Secretary

U.S. Geological Survey
Marcia K. McNutt, Director

U.S. Geological Survey, Reston, Virginia: 2011

For more information on the USGS—the Federal source for science about the Earth, its natural and living resources, natural hazards, and the environment, visit http://www.usgs.gov or call 1–888–ASK–USGS.

For an overview of USGS information products, including maps, imagery, and publications, visit http://www.usgs.gov/pubprod

To order this and other USGS information products, visit http://store.usgs.gov

Suggested citation:
Heilweil, V.M. and Marston, T.M., 2011, Assessment of managed aquifer recharge from Sand Hollow Reservoir, Washington County, Utah, updated to conditions in 2010: U.S. Geological Survey Scientific Investigations Report 2011–5142, 39 p.

Contents

Figures

Tables

Conversion Factors, Datums, and Water-Quality Units

Inch/Pound to SI

Multiply	By	To obtain
Length		
inch (in.)	2.54	centimeter (cm)
foot (ft)	0.3048	meter (m)
mile (mi)	1.609	kilometer (km)
Area		
acre	4,047	square meter (m^2)
acre	0.004047	square kilometer (km^2)
square foot (ft^2)	929.0	square centimeter (cm^2)
square foot (ft^2)	0.09290	square meter (m^2)
square inch (in^2)	6.452	square centimeter (cm^2)
square mile (mi^2)	2.590	square kilometer (km^2)
Volume		
ounce, fluid (fl. oz)	0.02957	liter (L)
pint (pt)	0.4732	liter (L)
quart (qt)	0.9464	liter (L)
gallon (gal)	3.785	liter (L)
cubic foot (ft^3)	0.02832	cubic meter (m^3)
acre-foot (acre-ft)	1,233	cubic meter (m^3)
Flow rate		
acre-foot per day (acre-ft/d)	0.01427	cubic meter per second (m^3/s)
Mass		
pound, avoirdupois (lb)	0.4536	kilogram (kg)
Density		
pound per cubic foot (lb/ft^3)	16.02	kilogram per cubic meter (kg/m^3)
Hydraulic conductivity		
foot per day (ft/d)	0.3048	meter per day (m/d)

Temperature in degrees Celsius (°C) may be converted to degrees Fahrenheit (°F) as follows: °F=(1.8×°C)+32

Temperature in degrees Fahrenheit (°F) may be converted to degrees Celsius (°C) as follows: °C=(°F-32)/1.8

Vertical coordinate information is referenced to the North American Vertical Datum of 1988 (NAVD 88).

Horizontal coordinate information is referenced to the North American Datum of 1983 (NAD 83).

Altitude, as used in this report, refers to distance above the vertical datum.

Specific conductance is given in microsiemens per centimeter at 25 degrees Celsius (µS/cm at 25°C).

Concentrations of chemical constituents in water are given either in milligrams per liter (mg/L) or micrograms per liter (µg/L).

Tritium units (TU) are used to report tritium concentration, where one TU equals tritium concentration in picoCuries per liter divided by 3.22.

Chlorofluorocarbon concentrations are reported as picomoles per kilogram (pmol/kg).

Sulfur hexafluoride concentrations are reported as femtomoles per kilogram (fmol/kg).

Acronyms and Abbreviations

CFCs	Chlorofluorocarbons
Cl	Chloride
Cl:Br	Chloride to bromide ratios
DO	Dissolved oxygen
DOC	Dissolved organic carbon
R_2	Coefficient of determination
SF_6	Sulfur hexafluoride
SHSP	Sand Hollow State Park
TDG	Total dissolved gas
TU	Tritium units
USGS	U.S. Geological Survey
WCWCD	Washington County Water Conservancy District

Assessment of Managed Aquifer Recharge from Sand Hollow Reservoir, Washington County, Utah, Updated to Conditions in 2010

By Victor M. Heilweil and Thomas M. Marston

Abstract

Sand Hollow Reservoir in Washington County, Utah, was completed in March 2002 and is operated primarily for managed aquifer recharge by the Washington County Water Conservancy District. From 2002 through 2009, total surface-water diversions of about 154,000 acre-ft to Sand Hollow Reservoir have allowed it to remain nearly full since 2006. Groundwater levels in monitoring wells near the reservoir rose through 2006 and have fluctuated more recently because of variations in reservoir water-level altitude and nearby pumping from production wells. Between 2004 and 2009, a total of about 13,000 acre-ft of groundwater has been withdrawn by these wells for municipal supply. In addition, a total of about 14,000 acre-ft of shallow seepage was captured by French drains adjacent to the North and West Dams and used for municipal supply, irrigation, or returned to the reservoir.

From 2002 through 2009, about 86,000 acre-ft of water seeped beneath the reservoir to recharge the underlying Navajo Sandstone aquifer. Water-quality sampling was conducted at various monitoring wells in Sand Hollow to evaluate the timing and location of reservoir recharge moving through the aquifer. Tracers of reservoir recharge include major and minor dissolved inorganic ions, tritium, dissolved organic carbon, chlorofluorocarbons, sulfur hexafluoride, and noble gases. By 2010, this recharge arrived at monitoring wells within about 1,000 feet of the reservoir.

Introduction

Sand Hollow is a 20 mi² basin located in the southeastern part of Washington County, Utah, about 10 mi northeast of St. George (fig. 1). It is part of the Virgin River drainage of the Lower Colorado River Basin and the upper Mohave Desert ecosystem. Altitudes range from 3,000 to 5,000 ft. Sand Hollow is underlain by Navajo Sandstone that is either exposed at the surface or covered by a veneer of soil or surface-flood basalts (Hurlow, 1998). The stratigraphic thickness of the Navajo Sandstone ranges from a few hundred feet to more than 1,200 ft. Sand Hollow Reservoir (fig. 1) was constructed in 2002 to provide surface-water storage and managed aquifer recharge to the underlying Navajo Sandstone. The reservoir is an off-stream facility that receives water from the Virgin River, diverted near the town of Virgin, Utah. The reservoir is impounded by two dams. The North Dam is about 3,200 ft long and is oriented east-southeast to west-northwest; the West Dam is about 6,000 ft long and is oriented north-northeast to south-southwest (fig. 2). At full capacity, the reservoir contains about 60,000 acre-ft of water and covers about 1,400 acres.

Sand Hollow has been the subject of interdisciplinary, cooperative investigations of groundwater hydrology and geochemistry since 1999. Previous Sand Hollow reports document pre-reservoir vadose-zone and groundwater conditions prior to March 2002 (Heilweil and Solomon, 2004; Heilweil and others, 2006; Heilweil and others, 2007; Heilweil and McKinney, 2007; Heilweil and others, 2009), pond and trench infiltration studies adjacent to the reservoir (Heilweil and others, 2004; Heilweil and Watt, 2011), and post-reservoir groundwater conditions, water budgets, and estimates of groundwater recharge from the reservoir from March 2002 through December 2007 (Heilweil and others, 2005; Heilweil and Susong, 2007; Heilweil and others, 2009). These reports also contain monitoring-well and production-well completion information, as well as historical water-quality and precipitation data. The objectives of this report are to present and interpret (1) groundwater levels, reservoir altitude, well withdrawals, drain discharge, meteorologic data, reservoir water temperature, and inflows/outflows from March 2002 through December 2009 for estimating monthly amounts of managed aquifer recharge from Sand Hollow Reservoir to the Navajo Sandstone, and (2) groundwater and surface water chemical data collected prior to the construction of the reservoir through March 2010 for evaluating groundwater flow paths and travel times of this managed aquifer recharge. This study is a cooperative effort by the Washington County Water Conservancy District (WCWCD) and the U.S. Geological Survey (USGS). Support for this work was provided by both the USGS and the WCWCD.

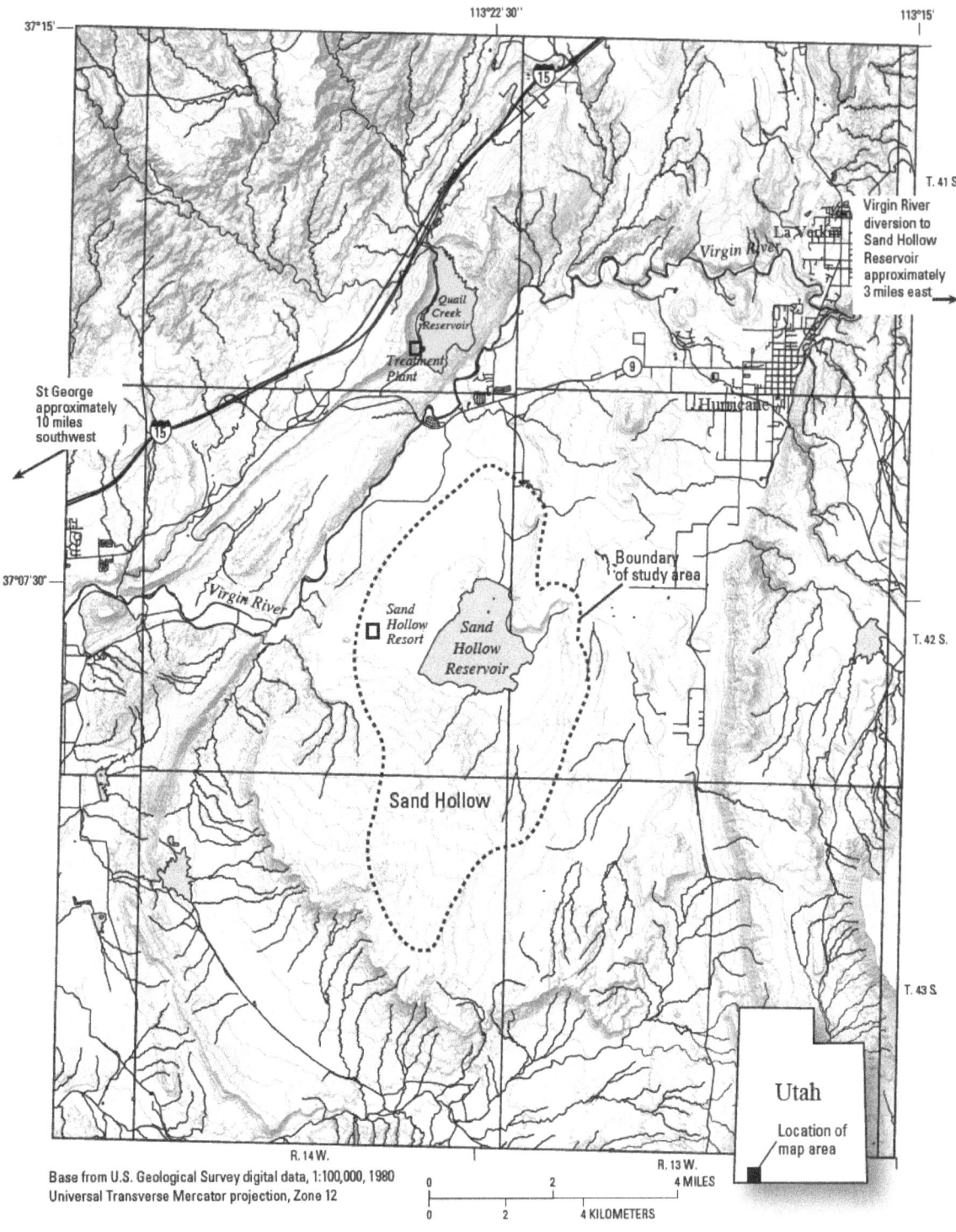

Figure 1. Location of the Sand Hollow study area, Washington County, Utah.

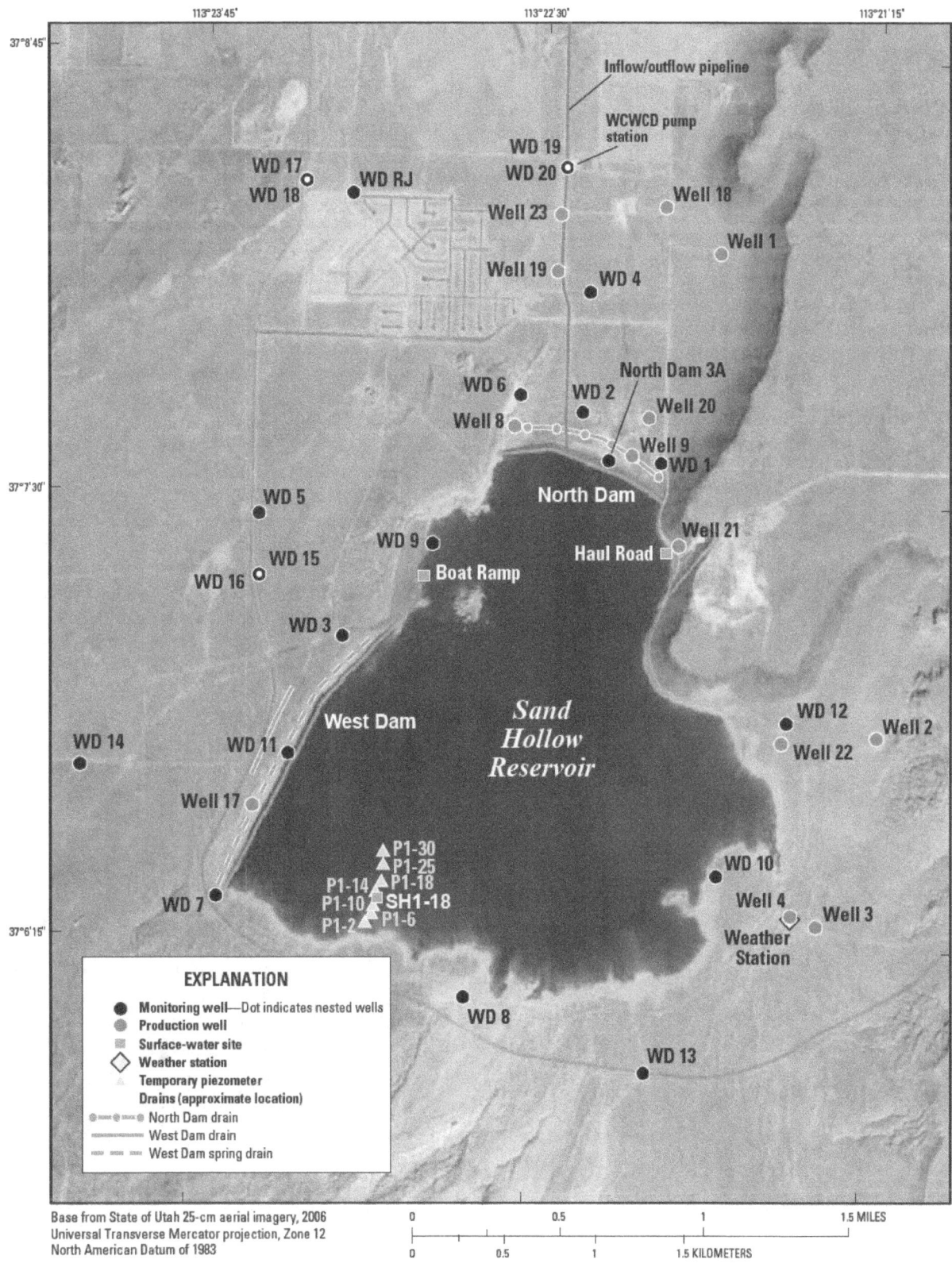

Figure 2. Location of wells, the weather station, drains, temporary piezometers, and surface-water sampling sites in Sand Hollow, Utah.

Assessment of Managed Aquifer Recharge from Sand Hollow Reservoir

Many different types of data have been collected to investigate recharge processes, to quantify recharge from Sand Hollow Reservoir, and to evaluate hydraulic and geochemical changes in the underlying Navajo Sandstone aquifer. These data include production-well withdrawals near the reservoir, amounts of pumpage from drains capturing shallow ground-water discharge adjacent to the reservoir, reservoir and moni-toring-well water levels, meteorological parameters, reservoir water temperatures, and inflows and outflows through the pipeline connecting Sand Hollow Reservoir with the Virgin River and the Quail Creek Reservoir and Treatment plant.

Data Collection Methods and Results

Data collection methods are described in detail in Heilweil and others (2005) and briefly summarized in the following sections.

Production-Well Withdrawals

The WCWCD has 13 production wells completed in the Navajo Sandstone available to capture both pre-existing groundwater (natural recharge) in Sand Hollow and recharge from Sand Hollow Reservoir (fig. 2). The WCWCD and other water users have withdrawn natural recharge in Sand Hollow for many years. The WCWCD groundwater withdrawals are recorded monthly from in-line magnetic flow meters installed at each well. Since August 2004, monthly withdrawals by the WCWCD have generally exceeded 150 acre-ft per month, except for several months during the winters of 2004–05, 2005–06, and 2008–09 (fig. 3). The majority of this pumping has been from Wells 8 and 9, both located adjacent to the North Dam. From 2004 through 2006, there were minimal withdrawals from these wells during the winter. Since 2006, withdrawals have been more constant year-round. Combined monthly pumping from these two production wells averaged about 230 acre-ft from March 2006 through December 2009. Smaller amounts have been withdrawn from Wells 1, 2, 17, and 21. A total of about 13,000 acre-ft were pumped from the WCWCD production wells from January 2004 through December 2009. Through 2009, withdrawals by the WCWCD at Sand Hollow have been permitted by the Utah Division of Water Rights as natural recharge in Sand Hollow. These with-drawals are governed by different water rights than recharge from Sand Hollow Reservoir; withdrawal rights for this artifi-cial recharge have not yet been exercised.

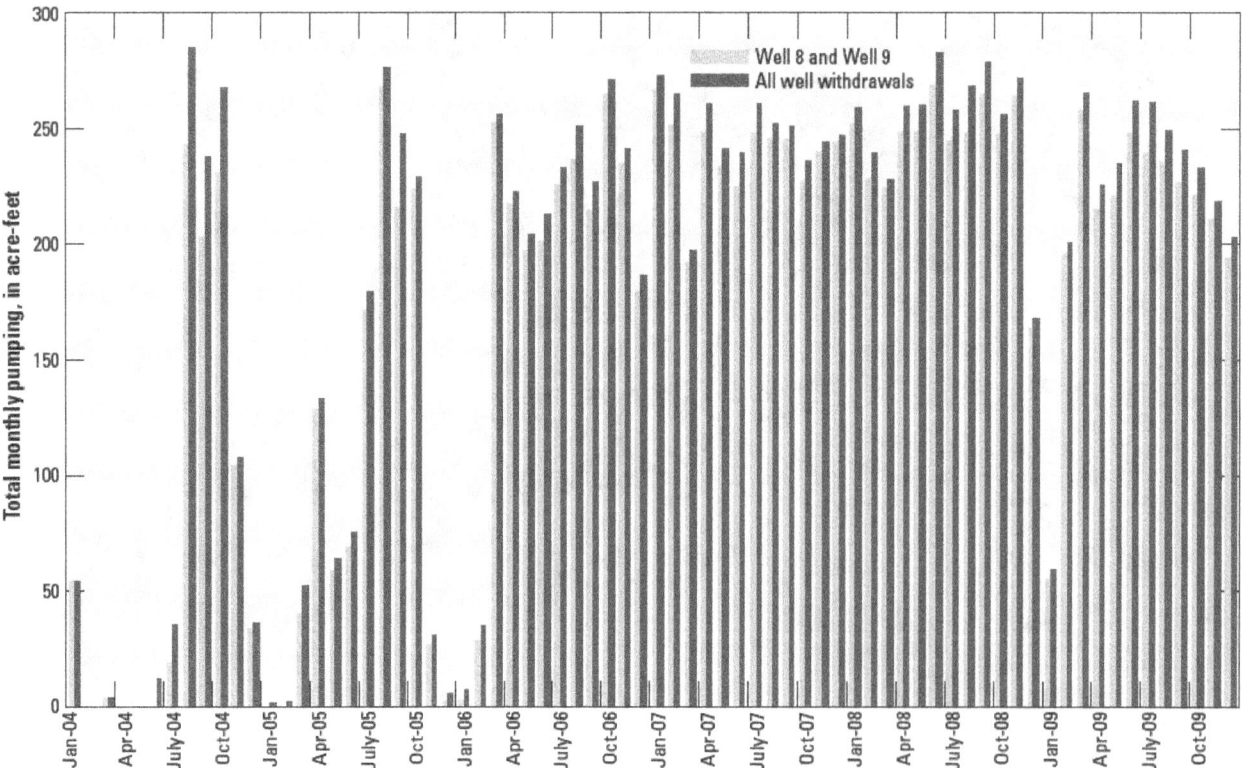

Figure 3.   Washington County Water Conservancy District production-well withdrawals in Sand Hollow, Utah, 2004–09.

Drain Discharge

Because of the steep gradients associated with the hydraulic connection between the reservoir and the underlying Navajo aquifer, some land-surface areas downgradient of the North and West Dams became saturated following construction of the reservoir. In response, three French drains (North Dam drain, West Dam drain, and West Dam Spring drain) were constructed for capturing this shallow groundwater. The North Dam drain was constructed in 2003 and is about 3,000 ft long, parallel to the North Dam (fig. 2). After scraping away surficial soils, it was excavated into the Navajo Sandstone with a trenching machine to a depth of about 20 ft and was built to capture shallow seepage beneath the North Dam. The West Dam drain, constructed in 2005, is about 1,500 ft long and located about 500 ft west of the center of the West Dam. It was excavated into unconsolidated soils with a backhoe to a depth of about 10 ft. The West Dam Spring drain was constructed in 2006 and is about 6,000 ft long, parallel to the West Dam. It is situated closer to the West Dam than the West Dam drain. Similar to the North Dam drain, unconsolidated soils were first removed before it was excavated to a depth of about 20 ft into the Navajo Sandstone with a trenching machine.

Amounts of discharge pumped from these drains are measured with a Tigermag totalizing flow meter, manufactured by Sparling Instruments in El Monte, California. Discharge to the North Dam drain has been pumped relatively consistently since September 2003 (fig. 4). About 4,800 acre-ft were pumped from the North Dam drain between 2003 and 2009. Initially, all of this water was returned to the reservoir, but since 2007, the majority of it, along with an additional 3,300 acre-ft of outflow from Sand Hollow Reservoir, has been used by Sand Hollow Resort (fig. 2) to meet increasing summer demand for irrigation. About 800 acre-ft of water were pumped from the West Dam drain back into the reservoir from 2005 through 2009. Beginning in October 2006, pumping of discharge from the West Dam Spring drain was initiated and largely has alleviated the need for pumping of the West Dam drain; from 2006 through 2009, about 8,500 acre-ft were pumped from the West Dam Spring drain for municipal use. While discharge from the West Dam Spring drain likely does not vary greatly, pumping from this drain is highly variable and dependent on other Sand Hollow Reservoir management and operations factors.

Groundwater-Level Data and Reservoir Altitude

Groundwater levels from a monitoring well network are used to document changes in the potentiometric surface associated with recharge from Sand Hollow Reservoir. The WCWCD measures water levels monthly in 21 monitoring wells completed in the Navajo Sandstone (table 1, fig. 2). These wells were constructed with either 1- or 2-inch diameter pvc casing, with perforations along the bottom 5- to 20-ft

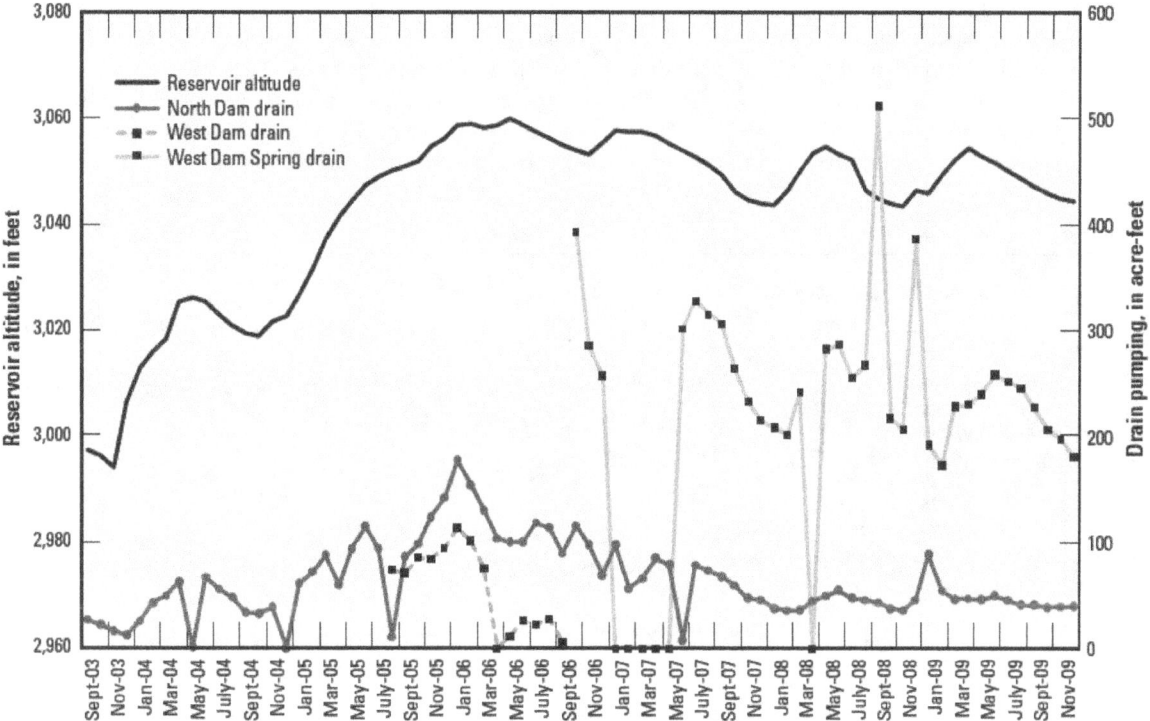

Figure 4.   Monthly reported reservoir altitude and discharge from the North Dam drain, West Dam drain, and West Dam Spring drain in Sand Hollow, Utah, 2003–09.

length of the casing. Three locations have nested pairs of water district (WD) monitoring wells: WD 15 and WD 16, WD 17 and WD 18, and WD 19 and WD 20. Wells measured by the WCWCD have had annual independent check measurements performed by USGS personnel for quality assurance to ensure accuracy of equipment (electric tape water-level indicators). In addition, daily reservoir water-level altitude (stage) was recorded using a pressure transducer installed by the WCWCD in the reservoir along the North Dam from January 2005 through December 2009. Because of periods of poor transducer reliability (periods when the transducer was not in calibration) from 2005 through 2007, daily reservoir altitude was interpolated on the basis of monthly measurements recorded at the boat ramp by WCWCD and Sand Hollow State Park (SHSP) personnel, and then correlated with trends from the transducer data. From 2008 through 2009, the transducer data were within 0.25 ft of the intermittent boat ramp measurements; thus, for this more-recent period, the daily reservoir stage recorded by the pressure transducer was used.

Recently measured (January 2008 through December 2009) and previously reported (Heilweil and others, 2005; Heilweil and Susong, 2007; Heilweil and others, 2009) groundwater levels and reservoir water-level altitude are shown in figure 5. The reservoir altitude rose from about 2,980 ft at the beginning of March 2002 to a maximum of about 3,060 ft in May 2006, when the reservoir was first filled to capacity. The reservoir altitude receded to about 3,040 ft in December 2007, and then fluctuated between about 3,040 and 3,060 ft during 2008 and 2009. The topographically lowest part of the reservoir bottom, adjacent to the North Dam, was the first region to be inundated with surface water in 2002 and 2003. As the reservoir continued to fill from 2004 through 2006, the extent of surface water increased toward the south in a line roughly perpendicular to the West Dam. Therefore, the monitoring wells nearest the northern side of the reservoir

were the first to show water-level responses and hydraulic connection with the reservoir. Water levels in WD 1, 2, 6, and 9 rose rapidly beginning in the spring of 2002. Water levels in WD 3 and WD 11, located farther south along the western side of the reservoir, began to rise rapidly in November 2002 and January 2003, respectively. Water levels in WD 10 and WD 12 (located on the eastern side of the reservoir) and in WD 7 and WD 8 (located along the southern side of the reservoir) began rising in the latter half of 2003. From 2004 through 2009, measured water levels in monitoring wells closer to the reservoir (WD 3, 7, 8, 9, 10, 11, 12) generally fluctuated with reservoir altitude. Exceptions to this were water levels in the monitoring wells near the North Dam (WD 1, 2, 6), which all had sharp rises during the winters of 2003–04, 2004–05, 2005–06, and 2008–09 associated with the temporary cessation of pumping at nearby wells 8 and 9. At monitoring wells farther from the reservoir (WD 4, 5, RJ, 13, 14), water levels generally displayed a more subdued rise in response to recharge beneath the reservoir. Water levels have been measured only since May 2009 in WD 15, 16, 17, 18, 19, and 20.

On the basis of water-level measurements in 21 monitoring wells, altitudes of the groundwater table near Sand Hollow Reservoir during August 2009 ranged from 2,907 to 3,041 ft (green points on fig. 6). The reservoir altitude during this same time period was about 3,049 ft. The lines on figure 6 show the estimated potentiometric contours in the aquifer (lines of equal groundwater-level altitude) and the arrows indicate the generalized directions of horizontal groundwater flow away from the reservoir. Horizontal hydraulic gradients, calculated by dividing the difference in water-level altitudes between two points by the distance separating these locations, indicate the direction of groundwater flow. The steepest horizontal hydraulic gradients are located beneath the North and West Dams and generally decline with increasing distance from the reservoir. For example, the horizontal hydraulic gradient between Sand

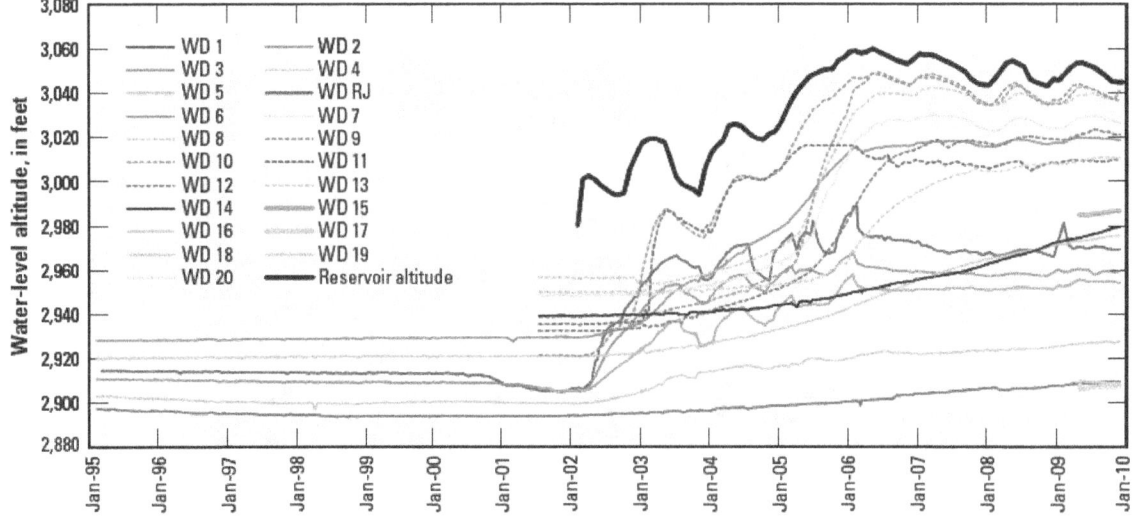

Figure 5. Relation between water level in monitoring wells and reservoir altitude, Sand Hollow, Utah, 1995–2009.

Table 1. Records of selected wells in Sand Hollow, Utah.

[Well name: Refer to figure 2. Primary use of site: W, withdrawal; O, observation. Casing finish: F, sand with perforations; S, screen; X, open hole. —, no data available; E, estimated]

Well name	Well location	Site ID	Latitude	Longitude	Altitude of land surface (feet)	Year drilled	Primary use of site	Depth of hole (feet)	Borehole diameter (inches)	Casing diameter (inches)	Casing bottom (feet)	Casing finish (feet)
North Dam 3A	(C-42-14)13dcd-2	370738113221701	37.126919	-113.37081	2,989	2001	O	26	3	1	25.7	F 15–25
WD 1	(C-42-14)13ddd-1	370737113220101	37.126852	-113.36749	2,998	1995	O	110	3	1	110	F 100–110
WD 2	(C-42-14)13dca-1	370746113222301	37.129141	-113.37247	2,987	1995	O	104	3	1	104	F 94–104
WD 3	(C-42-14)23daa-1	370706113231001	37.118385	-113.38695	3,025	1995	O	164	3	1	164	F 144–164
WD 4	(C-42-14)13acd-2	370805113221701	37.134791	-113.37215	2,960	1995	O	90	3	1	90	F 80–90
WD 5	(C-42-14)23abc-1	370727113233001	37.124063	-113.39227	2,991	1995	O	160	3	1	160	F 150–160
WD RJ	(C-42-14)14aad-1	370821113231001	37.139177	-113.38686	2,949	1995	O	205	3	1	205	F 195–205
WD 6	(C-42-14)13cda-1	370746113223301	37.12990	-113.37633	3,001	2001	O	96	3.8	2	96	F 90.8–95.8
WD 7	(C-42-14)26bdd-1	370622113233901	37.106074	-113.39438	3,067	2001	O	139.6	3.8	2	131	F 124.8–129.8
WD 8	(C-42-14)25cdb-1	370604113224601	37.10160	-113.37903	3,064	2001	O	130	3.8	2	118.7	F 113.5–118.5
WD 9	(C-42-14)24bcd-1	370655113224901	37.122835	-113.38152	3,064	2001	O	155	3.8	2	155	F 149.8–154.8
WD 10	(C-42-13)30bcd-1	370624113213801	37.107535	-113.36347	3,058	2001	O	133.8	3.8	2	122	F 116.8–121.9
WD 11	(C-42-14)23ddc-1	370646113231901	37.112835	-113.39015	3,014	2001	O	112	3.8	2	98.5	F 93.5–98.5
WD 12	(C-42-13)19cdb-1	370654113213501	37.114777	-113.35928	3,081	1999	O	164.6	3.8	2	155.5	F 150.3–155.3
WD 13	(C-42-14)25ddd-1	370554113220601	37.098213	-113.36777	3,081	—	O	250 E	8 E	6 E	—	—
WD 14	(C-42-14)26bbb-1	370638113240201	37.112074	-113.40289	3,021	1970	O	645	10	8	645	X 5–645
WD 15	(C-42-14)23acc-1	370716113233201	37.12119	-113.39217	3,000	2008	O	65	3.8	2	59	F 38.2–58.2
WD 16	(C-42-14)23acc-2	370716113233202	37.121163	-113.39217	3,000	2008	O	308	3.8	2	303	F 281.6–301.6
WD 17	(C-42-14)14abd-1	370822113232401	37.139694	-113.38975	2,993	2008	O	116	3.8	2	102.5	F 82.5–102.5
WD 18	(C-42-14)14abd-2	370822113232402	37.139727	-113.38975	2,993	2008	O	187	3.8	2	181.5	F 161.5–181.5
WD 19	(C-42-14)12dcc-1	370826113222601	37.14060	-113.37378	2,947	2008	O	70	3.8	2	65	F 43.8–63.8
WD 20	(C-42-14)12dcc-2	370826113222602	37.140619	-113.37378	2,947	2008	O	297	3.8	2	292	F 271.2–291.2
Well 1	(C-42-13)18bcb-2	370812113214801	37.13673	-113.36404	2,970	2003	W	1,005	24	16	1,000	S 120–1,000
Well 2	(C-42-13)19dcb-1	370651113211301	37.114177	-113.35367	3,159	2002	W	965	24	16	900	S 135–855
Well 3	(C-42-13)30bdd-1	370619113212601	37.10528	-113.35719	3,080	1994	W	590	16	13	52	X 52–590
Well 4	(C-42-13)30bdc-1	370621113213201	37.105749	-113.35877	3,070	1983	W	600	18	12	—	—
Well 8	(C-42-14)13cdd-1	370742113223601	37.128418	-113.37666	3,030	2001	W	624	24	16	624	S 144–624
Well 9	(C-42-14)13ddc-1	370738113220901	37.127162	-113.36935	2,990	2000	W	1,210	24	16	1,210	S 104–1,140
Well 17	(C-42-14)26abb-1	370638113233401	37.110377	-113.39226	3,027	2000	W	608	24	16	600	S 140–580
Well 18	(C-42-14)13aad-1	370821113220101	37.138853	-113.36749	2,951	2007	W	660	26	18	650	S 100–640
Well 19	(C-42-14)13acc-1	370809113222701	37.135734	-113.37421	2,970	2006	W	645	26	18	640	S 130–630
Well 20	(C-42-14)13dda-1	370744113220601	37.128963	-113.36832	2,980	2007	W	575	26	18	560	S 100–540
Well 21	(C-42-13)19dcb-1	370723113215901	37.122987	-113.36626	3,080	2006	W	605	26	18	600	S 100–580
Well 22	(C-42-13)19cdc-1	370650113213401	37.113832	-113.35955	3,075	2007	W	610	26	18	600	S 100–580
Well 23	(C-42-14)13abc-1	370818113222701	37.138419	-113.37407	2,967	2006	W	650	26	18	640	S 120–600

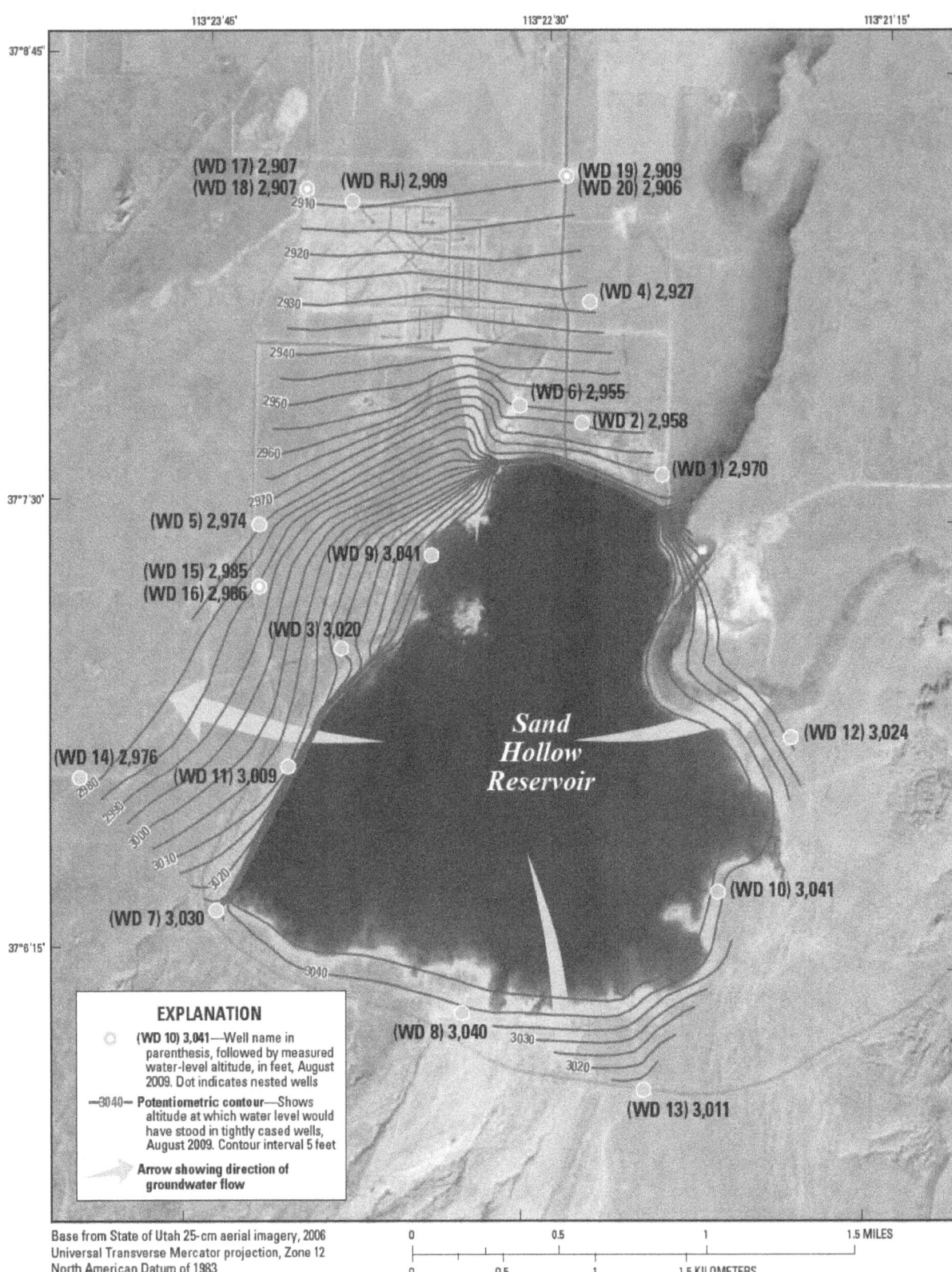

Figure 6.   Potentiometric surface of the Navajo aquifer in August 2009, Sand Hollow, Utah.

Table 2. Monthly reservoir data, evaporation, and groundwater recharge from Sand Hollow Reservoir, Utah, 2002–09.

[Reservoir altitude and reservoir storage: value is from the last day of each month; reservoir surface area: value is an average of the daily values for each month; 2σ, 2 sigma; —, no data available]

Month	Reservoir altitude (feet)	Reservoir storage (acre-feet)	Monthly pump station inflow (+) or outflow (-) (acre-feet)	Monthly drain and spring return flow to reservoir (acre-feet)	Monthly outflow (-) to Sand Hollow Resort (acre-feet)	Monthly total inflow (+) or outflow (-) to/from reservoir (acre-feet)	Monthly reservoir storage change (acre-feet)	Reservoir surface area (acres)	Monthly evaporation rate (feet)[1]	Monthly evaporation (acre-feet)	Monthly precipitation (acre-feet)	Monthly groundwater recharge (acre-feet)	Monthly groundwater recharge uncertainty, 2σ (percent)	Monthly groundwater recharge uncertainty, 2σ (acre-feet)	Groundwater recharge rate (feet/day)
Mar.-02	3,001	3,090	6,620	0	0	6,620	3,090	260	0.24	60	—	3,470	6.7	232	0.430
Apr.-02	3,003	3,500	3,690	0	0	3,690	410	280	0.46	130	—	3,150	5.9	187	0.383
May-02	3,001	3,090	2,450	0	0	2,450	-410	260	0.68	180	—	2,680	6.6	176	0.330
June-02	2,999	2,480	0	0	0	0	-610	230	0.91	210	—	400	12.6	50	0.058
July-02	2,997	2,050	0	0	0	0	-430	210	0.90	190	—	240	13.1	31	0.040
Aug.-02	2,995	1,650	0	0	0	0	-400	180	0.81	150	—	250	12.7	32	0.044
Sept.-02	2,994	1,300	0	0	0	0	-350	140	0.47	70	—	280	11.7	33	0.070
Oct.-02	2,995	1,500	790	0	0	790	200	160	0.26	40	—	550	6.6	36	0.110
Nov.-02	3,006	4,220	3,590	0	0	3,590	2,720	320	0.11	30	—	840	7.2	61	0.090
Dec.-02	3,012	7,000	3,930	0	0	3,930	2,780	400	0.05	20	—	1,130	7.1	80	0.090
Jan.-03	3,017	9,760	4,580	0	0	4,580	2,760	590	0.09	50	—	1,770	7.0	123	0.097
Feb.-03	3,019	10,670	2,850	0	0	2,850	910	570	0.10	60	—	1,880	6.4	121	0.118
Mar.-03	3,020	10,930	1,930	0	0	1,930	260	580	0.24	140	—	1,530	6.5	99	0.085
Apr.-03	3,019	10,680	540	0	0	540	-250	570	0.37	210	—	580	9.4	55	0.034
May-03	3,018	9,930	0	0	0	0	-750	540	0.66	360	—	390	13.2	52	0.023
June-03	3,010	6,040	-3,120	0	0	-3,120	-3,890	390	0.89	350	—	420	8.4	35	0.036
July-03	3,002	3,200	-2,020	0	0	-2,020	-2,840	240	0.92	220	—	600	8.4	51	0.081
Aug.-03	2,999	2,540	0	30	0	30	-660	230	0.75	170	—	490	12.0	59	0.069
Sept.-03	2,997	2,100	0	20	0	20	-440	220	0.58	130	—	340	12.3	42	0.052
Oct.-03	2,996	1,850	0	20	0	20	-250	170	0.36	60	—	210	11.9	25	0.040
Nov.-03	2,994	1,560	0	20	0	20	-290	200	0.09	20	—	290	10.6	31	0.048
Dec.-03	3,007	4,700	3,590	10	0	3,600	3,140	330	0.06	20	—	440	7.4	32	0.043
Jan.-04	3,013	7,600	3,990	30	0	4,020	2,900	480	0.06	30	—	1,090	7.2	78	0.073
Feb.-04	3,016	8,840	2,320	40	0	2,360	1,240	600	0.08	40	—	1,080	6.9	74	0.064
Mar.-04	3,019	10,400	2,400	50	0	2,450	1,560	630	0.38	240	—	650	7.7	50	0.033
Apr.-04	3,025	15,070	5,620	60	0	5,680	4,670	750	0.42	310	—	700	7.6	53	0.031
May-04	3,026	15,830	2,050	0	0	2,050	760	780	0.72	560	—	730	8.6	63	0.030
June-04	3,025	14,400	0	70	0	70	-1,430	750	0.87	650	—	850	13.1	112	0.038
July-04	3,023	13,000	0	60	0	60	-1,400	680	0.94	640	—	820	13.1	108	0.039
Aug.-04	3,021	11,670	0	50	0	50	-1,330	680	0.78	530	—	850	12.8	109	0.040
Sept.-04	3,019	11,260	[2]600	30	0	630	-410	630	0.53	330	—	710	10.2	73	0.038
Oct.-04	3,019	11,040	[2]630	30	0	660	-220	610	0.25	150	—	730	8.4	61	0.039
Nov.-04	3,022	12,650	[2]2,300	40	0	2,340	1,610	630	0.10	70	—	660	7.3	48	0.035
Dec.-04	3,023	13,390	[2]1,400	0	0	1,400	740	670	0.06	40	—	620	7.0	43	0.030

Table 2. Monthly reservoir data, evaporation, and groundwater recharge from Sand Hollow Reservoir, Utah, 2002–09.—Continued

[Reservoir altitude and reservoir storage: value is from the last day of each month; reservoir surface area: value is an average of the daily values for each month; 2σ, 2 sigma; —, no data available]

Month	Reservoir altitude (feet)	Reservoir storage (acre-feet)	Monthly pump station inflow (+) or outflow (-) (acre-feet)	Monthly drain and spring return flow to reservoir (acre-feet)	Monthly outflow (-) to Sand Hollow Resort (acre-feet)	Monthly total inflow (+) or outflow (-) to/from reservoir (acre-feet)	Monthly reservoir storage change (acre-feet)	Reservoir surface area (acres)	Monthly evaporation rate (feet)[1]	Monthly evaporation (acre-feet)	Monthly precipitation (acre-feet)	Monthly groundwater recharge (acre-feet)	Monthly groundwater recharge uncertainty, 2σ (percent)	Monthly groundwater recharge uncertainty, 2σ (acre-feet)	Groundwater recharge rate (feet/day)
Jan.-05	3,027	16,200	[2]3,500	60	0	3,560	2,810	740	0.07	50	—	700	7.3	51	0.031
Feb.-05	3,032	20,280	[2]5,200	70	0	5,270	4,080	780	0.11	80	130	1,240	7.3	91	0.057
Mar.-05	3,037	25,030	6,530	90	0	6,620	4,750	880	0.24	210	100	1,760	7.4	130	0.065
Apr.-05	3,041	29,220	6,180	60	0	6,240	4,190	960	0.39	370	130	1,810	7.5	136	0.063
May-05	3,044	32,370	5,140	90	0	5,230	3,150	1,020	0.70	710	40	1,410	7.9	112	0.045
June-05	3,048	35,750	6,100	110	0	6,210	3,380	1,080	0.75	810	20	2,040	7.8	160	0.063
July-05	3,049	37,280	3,600	90	0	3,690	1,530	1,120	0.97	1,080	10	1,090	8.8	96	0.031
Aug.-05	3,050	38,670	3,390	80	0	3,470	1,390	1,140	0.75	850	40	1,270	8.5	108	0.036
Sept.-05	3,051	39,580	3,010	160	0	3,170	910	1,160	0.54	630	20	1,650	8.1	133	0.047
Oct.-05	[3]3,052	[3]40,960	2,960	180	0	3,140	[3]1,380	[3]1,190	0.28	330	60	1,490	7.6	113	0.040
Nov.-05	3,055	44,310	5,160	210	0	5,370	[3]3,350	1,230	0.11	140	40	1,920	7.2	138	0.052
Dec.-05	3,056	46,120	3,380	240	0	3,620	1,810	1,250	0.05	60	20	1,770	6.9	122	0.046
Jan.-06	3,059	49,590	4,660	290	0	4,950	3,470	1,290	0.08	100	10	1,390	7.3	101	0.035
Feb.-06	3,059	49,840	1,200	250	0	1,450	250	1,320	0.12	160	30	1,070	7.3	78	0.029
Mar.-06	3,058	48,700	60	210	0	270	-1,140	1,310	0.18	240	60	1,230	11.4	140	0.030
Apr.-06	3,059	49,450	2,060	100	0	2,160	750	1,300	0.45	580	40	870	8.7	76	0.022
May-06	3,060	51,280	3,650	110	0	3,760	1,830	1,320	0.76	1,000	0	930	8.7	81	0.023
June-06	3,059	49,520	10	130	0	140	-1,760	1,330	0.92	1,220	10	690	14.1	97	0.017
July-06	3,058	47,920	30	140	0	170	-1,600	1,310	0.88	1,160	30	640	14.1	90	0.016
Aug.-06	3,056	46,220	0	140	0	140	-1,700	1,280	0.80	1,020	0	820	13.8	113	0.021
Sept.-06	3,055	44,610	10	90	0	100	-1,610	1,260	0.52	650	10	1,070	12.8	137	0.028
Oct.-06	3,054	43,390	30	120	0	150	-1,220	1,230	0.22	270	30	1,130	11.6	132	0.030
Nov.-06	3,053	42,360	0	100	0	100	-1,030	1,220	0.07	90	0	1,040	10.8	112	0.028
Dec.-06	3,055	45,100	4,430	70	0	4,500	2,740	1,230	0.04	60	10	1,710	7.0	120	0.045
Jan.-07	3,058	48,230	4,190	100	0	4,290	3,130	1,270	0.05	60	10	1,110	7.2	80	0.028
Feb.-07	3,057	47,630	30	60	0	90	-600	1,290	0.13	170	30	550	11.9	65	0.015
Mar.-07	3,057	47,660	1,210	70	0	1,280	30	1,290	0.33	430	0	820	9.0	73	0.021
Apr.-07	3,057	46,720	50	80	0	130	-940	1,280	0.45	580	50	540	13.4	73	0.014
May-07	3,055	44,880	0	0	-110	-110	-1,840	1,220	0.74	900	0	830	13.3	110	0.022
June-07	3,054	43,390	0	0	-220	-220	-1,490	1,240	0.93	1,150	0	120	14.4	17	0.003
July-07	3,053	41,740	120	0	-200	-80	-1,650	1,210	0.92	1,110	110	560	13.5	76	0.015
Aug.-07	3,051	40,040	60	0	-210	-150	-1,700	1,180	0.81	960	60	650	13.3	87	0.018
Sept.-07	3,050	38,040	[4]-750	0	-210	[5]-910	-2,000	1,160	0.57	660	80	510	10.8	55	0.015

Table 2.   Monthly reservoir data, evaporation, and groundwater recharge from Sand Hollow Reservoir, Utah, 2002–09.—Continued

[Reservoir altitude and reservoir storage: value is from the last day of each month; reservoir surface area: value is an average of the daily values for each month; 2σ, 2 sigma; —, no data available]

Month	Reservoir altitude (feet)	Reservoir storage (acre-feet)	Monthly pump station inflow (+) or outflow (-) (acre-feet)	Monthly drain and spring return flow to reservoir (acre-feet)	Monthly outflow (-) to Sand Hollow Resort (acre-feet)	Monthly total inflow (+) or outflow (-) to/from reservoir (acre-feet)	Monthly reservoir storage change (acre-feet)	Reservoir surface area (acres)	Monthly evaporation rate (feet)[1]	Monthly evaporation (acre-feet)	Monthly precipitation (acre-feet)	Monthly groundwater recharge (acre-feet)	Monthly groundwater recharge uncertainty, 2σ (percent)	Monthly groundwater recharge uncertainty, 2σ (acre-feet)	Groundwater recharge rate (feet/day)
Oct.-07	3,046	34,280	[4]-2,660	0	-120	[5]-2,780	-3,760	1,120	0.32	360	0	620	8.6	53	0.018
Nov.-07	3,045	32,480	[4]-750	0	-100	[5]-850	-1,800	1,060	0.16	170	100	880	9.3	82	0.028
Dec.-07	3,044	31,680	90	10	0	100	-800	1,040	0.05	50	90	940	10.0	94	0.029
Jan.-08	3,044	31,470	0	20	0	20	-210	1,030	0.06	60	50	220	11.9	26	0.007
Feb.-08	3,046	34,490	3,240	20	0	3,260	3,020	1,050	0.13	140	100	200	7.7	15	0.007
Mar.-08	3,050	38,460	4,420	0	-70	4,350	3,970	1,110	0.29	320	10	70	7.8	5	0.002
Apr.-08	3,053	42,670	4,950	0	-160	4,790	4,210	1,180	0.45	530	0	50	8.0	4	0.001
May-08	3,055	44,410	3,260	0	-120	3,140	1,740	1,230	0.61	750	50	700	8.5	59	0.018
June-08	3,053	42,540	0	0	-220	-220	-1,870	1,230	0.93	1,140	10	520	13.8	72	0.014
July-08	3,052	41,080	0	0	-180	-180	-1,460	1,180	0.95	1,120	110	270	14.2	38	0.007
Aug.-08	3,047	34,600	[6]-5,000	0	-180	-5,180	-6,480	1,140	0.82	940	10	370	8.7	32	0.010
Sept.-08	3,045	32,960	0	0	-140	-140	-1,640	1,070	0.61	650	20	870	12.8	111	0.027
Oct.-08	3,044	31,890	0	0	-70	-70	-1,070	1,050	0.36	370	60	690	12.5	86	0.021
Nov.-08	3,043	31,160	0	0	-10	-10	-730	1,040	0.16	160	80	640	11.6	75	0.021
Dec.-08	3,046	34,490	4,100	40	0	4,140	3,330	1,050	0.06	60	50	800	7.4	59	0.025
Jan.-09	3,046	33,830	0	70	0	70	-660	1,080	0.09	100	50	680	11.2	76	0.020
Feb.-09	3,049	37,770	4,630	50	0	4,680	3,940	1,110	0.14	150	60	650	7.5	49	0.021
Mar.-09	3,052	41,320	4,800	0	-30	4,770	3,550	1,190	0.30	360	0	860	7.7	66	0.023
Apr.-09	3,055	44,030	3,920	0	-70	3,850	2,710	1,220	0.44	530	20	630	8.0	50	0.017
May-09	3,053	42,180	180	10	-170	20	-1,850	1,220	0.78	950	0	920	12.9	119	0.024
June-09	3,052	40,600	210	0	-130	80	-1,580	1,190	0.73	870	10	800	12.9	103	0.022
July-09	3,050	38,700	220	0	-170	50	-1,900	1,170	0.96	1,120	10	840	13.1	110	0.023
Aug.-09	3,049	36,960	210	0	-150	60	-1,740	1,140	0.80	910	0	890	12.8	114	0.025
Sept.-09	3,047	35,380	200	0	-150	50	-1,580	1,110	0.58	650	0	980	12.3	120	0.029
Oct.-09	3,046	33,940	200	10	-80	130	-1,440	1,090	0.30	320	[7]0	1,320	11.1	147	0.039
Nov.-09	3,045	32,960	180	10	-20	170	-980	1,070	0.16	170	[7]0	1,050	10.6	111	0.033
Dec.-09	3,044	32,320	200	40	0	240	-640	1,050	0.05	60	100	920	9.6	88	0.028
Total	NMF	NMF	NMF	NMF	NMF	153,610	NMF	NMF	NMF	37,360	NMF	86,230	NMF	NMF	NMF

[1]Monthly evaporation rate from February 2007 through December 2009 was calculated with a correction factor to account for higher solar radiation measurements with new instrument

[2]Because of problems with monitoring equipment, inflows from September 2004 through February 2005 are estimated based on previous inflow history and changes in reservoir altitude

[3]Revised value based on refined reservoir altitude estimate for October 2005

[4]Monthly pump station outflow was increased from amount reported in Heilweil and others (2009) based on reservoir altitude relations

[5]Monthly total outflow was increased from amount reported by WCWCD based on reservoir altitude relations

[6]Monthly pump station outflow was increased from previously reported amount (Heilweil and others, 2009) based on reservoir altitude relations

[7]Sand Hollow rain gauge not functioning; values of 0 based on lack of precipitation from St George precipitation station #427516

Hollow Reservoir (reservoir altitude of 3,049 ft) and WD 1 (groundwater altitude of 2,970 ft) in 2009 was 0.120 foot per foot (ft/ft), whereas the gradient between WD 4 (groundwater altitude of 2,927 ft) and WD 19 (2,909 ft) was 0.008 ft/ft. In 2009, the broader regional gradient between WD 9 (3,041 ft altitude) and WD RJ (2,909 ft altitude) was 0.021 ft/ft. In comparison, the hydraulic gradient between these same two wells in 2004 was 0.017 ft/ft (fig. 7 of Heilweil and others, 2005). The generalized hydraulic gradients, shown as blue arrows perpendicular to the potentiometric contours in figure 6, indicate that groundwater is moving laterally away from the reservoir in all directions. The flow paths shown to the south, east, and west of the reservoir, however, likely curve around toward the north farther away from the reservoir. Because of the erosional extent of the Navajo Sandstone to the south and west of Sand Hollow, along with the displacement along the Hurricane fault to the east, all natural groundwater discharge from Sand Hollow likely occurs to the north as seepage to the Virgin River (fig. 1; Heilweil and others, 2000).

Surface-Water Inflow to and Outflow from Sand Hollow Reservoir

Surface water is pumped into and flows out of Sand Hollow Reservoir through a 60-in. diameter pipeline that enters through the North Dam (fig. 2). Sand Hollow Reservoir is currently managed to maximize groundwater recharge and little surface water has been removed from the reservoir. Monthly surface-water inflow to and outflow from Sand Hollow Reservoir is shown in table 2. The "Monthly pump station inflow or outflow" column of this table is the amount of Sand Hollow Reservoir surface water coming from the Virgin River or going to the Quail Creek Reservoir Water Treatment Plant (fig. 1). These data were collected at the WCWCD pump station located about 1 mi north of the North Dam. Five turbines, each with Sparling Tigermag totalizing flow meters, are linked to a computer system that combines and records total daily discharge in gallons. The flow meters have electronic modules on which calibration diagnostics are performed monthly by the WCWCD. Each module is removed annually for factory recalibration.

A wetter period during 2004 and 2005 allowed the WCWCD to divert larger amounts of surface water to Sand Hollow Reservoir from the Virgin River and fill the reservoir to nearly full storage capacity by February 2006. Because 2006 was a dry year (only 2.1 in. of total rainfall recorded at Sand Hollow), very little water was diverted from the Virgin River to the reservoir. Larger amounts of precipitation from the latter half of 2007 through early 2009 allowed for increased diversions to the reservoir.

The "Monthly drain and spring return flow to reservoir" column of table 2 is the portion of discharge to the three drains that is pumped back into Sand Hollow Reservoir. The "Monthly outflow to Sand Hollow Resort" column is the amount of water required by the resort that cannot be met by discharge to the North Dam drain and is fulfilled by outflow

from Sand Hollow Reservoir. Therefore, the "Monthly total inflow or outflow to/from reservoir" column is a sum of the pump station inflow/outflow, the drain and spring return flow, and the outflow to Sand Hollow Resort (table 2).

The "Monthly pump station inflow or outflow" column is comparable to the "Total surface-water inflow or outflow" column in table 7 of Heilweil and others (2005), the "Monthly surface-water inflow or outflow" column in table 2 of Heilweil and Susong (2007), and the "Monthly net surface-water inflow/outflow" column in table 2 of Heilweil and others (2009). For this report, however, previously estimated (Heilweil and others, 2009) outflows were adjusted (increased) for September, October, and November of 2007 to better match the decline in the reservoir altitude. The previously published values for these 3 months (80, –580, and 100 acre-ft) have been revised to –750, –2,670, and –750 acre-ft. Although not reported, outflow rates must have peaked in mid-October 2007, as indicated by a 2-ft drop in reservoir altitude in less than 5 days.

Beginning with this report, both "Monthly drain and spring return flow to reservoir" and "Monthly outflow to Sand Hollow Resort" are included in calculations of total inflow to and outflow from the reservoir. These amounts are added to the "Monthly pump station inflow or outflow" and summed in the "Monthly total inflow or outflow to/from reservoir" column. Monthly total inflow/outflow amounts from March 2002 through December 2009 range from about –5,000 acre-ft to 6,600 acre-ft. Approximately 154,000 acre-ft of total net inflow have been pumped into Sand Hollow Reservoir from 2002 through 2009.

Meteorology Data

Meteorology data have been collected at a weather station (fig. 2) in Sand Hollow since January 1998. The weather station has been used for evaluating evaporation and precipitation, which are required for calculating monthly recharge from Sand Hollow Reservoir. Parameters measured include air temperature, wind speed, wind direction, precipitation, relative humidity, and incoming solar radiation. Instrumentation includes a temperature and relative humidity probe, a wind direction and speed monitor, a tipping bucket rain gage, and a solar radiometer. Sensors collect data every minute, and average hourly and daily values are computed and stored on a data logger (with the exception of precipitation, which is summed rather than averaged). The solar radiation and temperature data were used for calculating evaporation (using the Jenson and Haise method; see below). The other data were collected to permit calculations of evaporation using other methods.

From January 13, 1998, to December 30, 2009, daily average air temperature ranged from –2°C to 37°C. The coldest temperatures during the year typically occurred during December and January, when minimum air temperatures occasionally were below –8°C. The warmest temperatures were typically in July, when maximum air temperatures occasionally approached 45°C. Daily average solar radiation ranged from 34 to 840 calories per square centimeter per day. The minimum daily averages typically occur in December and January; the maximum daily averages typically occur in June and July.

Monthly precipitation has been recorded at Sand Hollow weather station continuously from January 1998 through December 2009, except for two periods when malfunctioning instrumentation resulted in data loss: December 26, 2008 to January 3, 2009, and September 28 to November 16, 2009. Precipitation amounts during these two periods were estimated based on data from the nearby St. George Southgate Golf Course weather station (#427516; http://www.wrcc.dri. edu/cgi-bin/cliMAIN.pl?ut7516). From January 1998 through December 2009, monthly precipitation ranged from 0 to more than 3.5 in. (fig. 7), and averaged about 0.5 in. Average annual precipitation during the 12-year period of 1998 through 2009 was 6.3 in. Annual precipitation exceeded 10 in. during 2004 and 2005, indicative of wetter-than-normal conditions in the Virgin River watershed.

Reservoir Water Temperature

Continuous water-temperature measurements were made in Sand Hollow Reservoir and used for evaluating effects of water viscosity changes on seepage rates beneath the reservoir. A string of five thermistors was installed in January 2003 in the deepest part of Sand Hollow Reservoir, about 300 ft from the North Dam. The initial thermistors were attached to a floating buoy at depths of about 0.3 ft (R1), 3 ft (R2), 10 ft (R3), 15 ft (R4), and 30 ft (R5). A sixth thermistor (R6) was added on May 6, 2008, at a depth of 50 ft (or at the bottom of the reservoir, if shallower). The thermistors are reported to have an accuracy of better than 0.5°C over the temperature range of 0 to 35°C. Water temperature from January 2003 through December 2009 has ranged from about 1 to 30°C. Both the previous (January 2003 through December 2007) and current (January 2008 through December 2009) temperature data are shown in figure 8. The following water temperatures were not recorded for periods exceeding 30 days during the current period because of problems with the thermistors and/or buoy: R1 at the 0.3-ft depth, January 1 to June 4, 2008; R2 at the 3-ft depth, March 11 to June 5, 2008; R3 at the 10-ft depth, March 11 to July 29, 2008; R4 at the 15-ft depth, March 11 to May 5, 2008, and August 27 to October 3, 2009; R5 at the 30-ft depth, June 5 to July 29, 2008; and R6 at the 50-ft depth, June 5 to July 29, 2008.

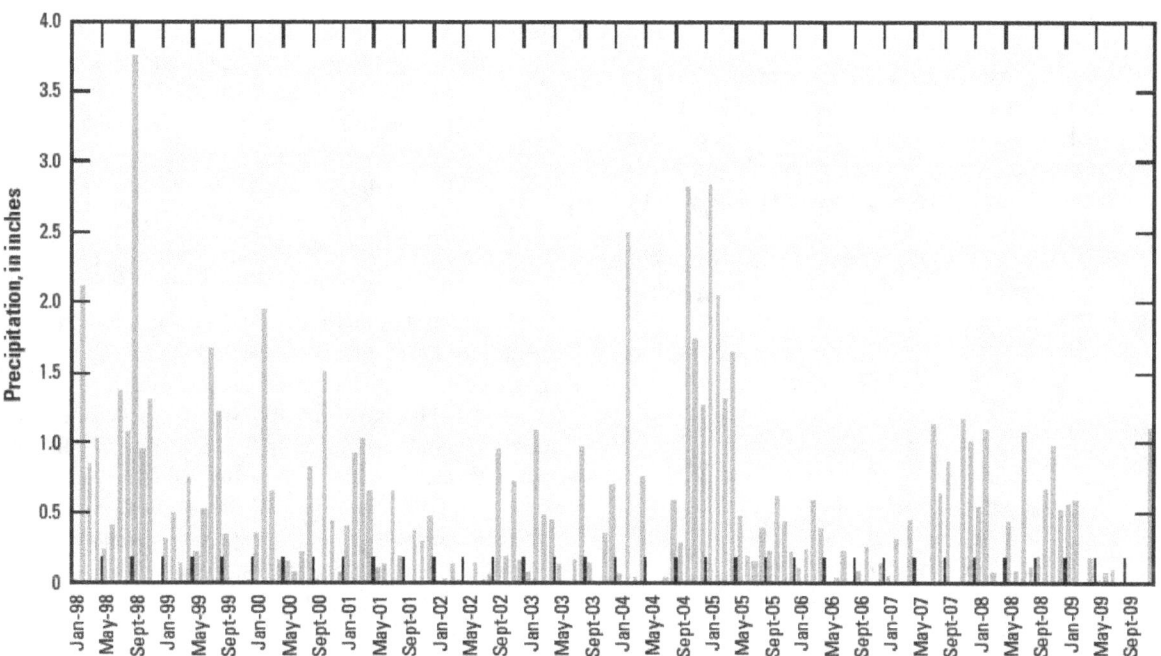

Figure 7.   Monthly precipitation at Sand Hollow, Utah, 1998–2009.

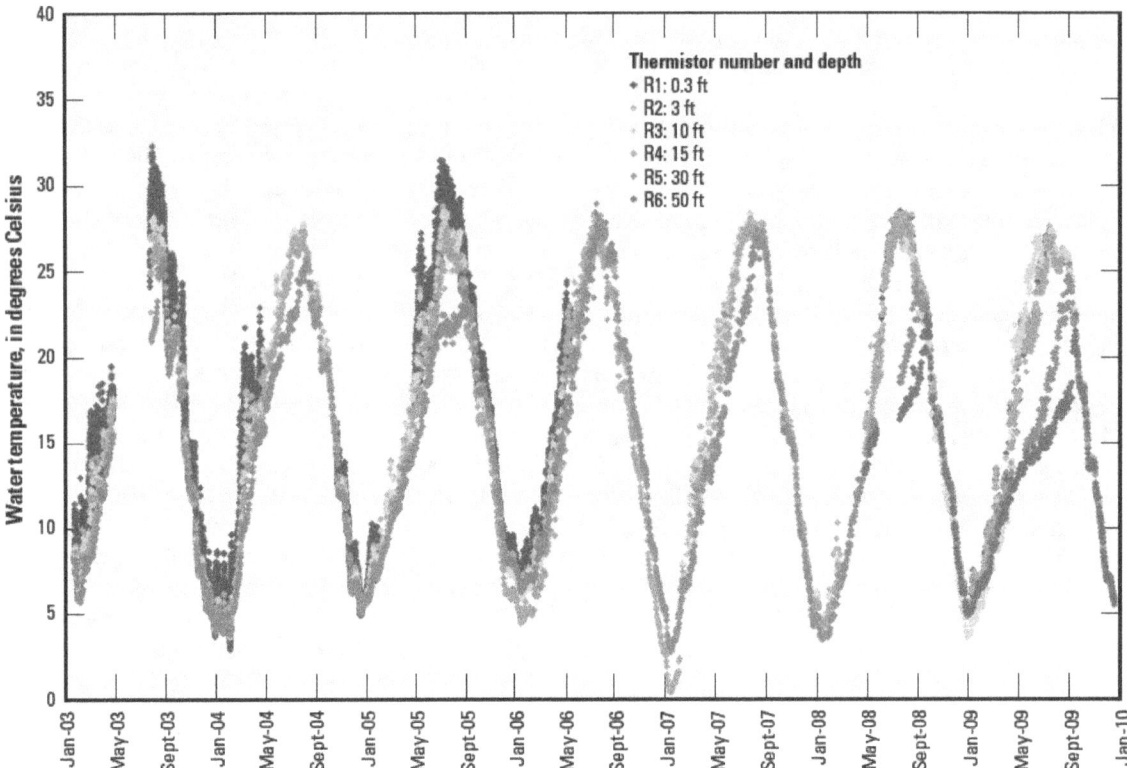

Figure 8.   Daily water temperature at various depths in Sand Hollow Reservoir, Utah, 2003–09.

Estimates of Managed Aquifer Recharge from Sand Hollow Reservoir

Substantial amounts of surface water from Sand Hollow Reservoir infiltrate through the underlying sediments to recharge the Navajo Sandstone aquifer. This recharge either is captured by production wells for municipal supply, or it moves northward through the aquifer to discharge as seepage to the Virgin River.

Recharge from Sand Hollow Reservoir is calculated with the following water-budget equation (modified from Heilweil and others, 2005):

$$R = I_{sw} + I_{DR} - O_{sw} + P \pm \Delta S - E \qquad (1)$$

where

R	is recharge,
I_{sw}	is surface-water inflow,
I_{Dr}	is drain and spring return flow,
O_{sw}	is surface-water outflow,
P	is the amount of precipitation falling directly on the reservoir,
ΔS	is change in surface-water storage, and
E	is evaporation.

All amounts for the variables of equation 1 are in acre-ft.

The following equation was developed to evaluate the uncertainty for each monthly recharge estimate:

$$CU = \Sigma[(|C_i|/\Sigma|C_i|)*U_i] \qquad (2)$$

where

CU	is the composite uncertainty fraction (2σ, two standard deviation)
$|C_i|$	is the absolute value of each component of the water budget (acre-ft),
$\Sigma|C_i|$	is the sum of absolute values of all the water-budget components (acre-ft), and
U_i	is the uncertainty fraction (2σ) for each individual water-budget component.

The smallest estimated uncertainty fraction is 0.05 (5 percent) for I_{sw}, I_{Dr}, and O_{sw} because these flows are recorded using calibrated inline flow meters. The estimated uncertainty fraction for P is higher, at 0.10 (10 percent), because it is an indirect measurement made on the basis of nearby meteorology station data. Similarly, the estimated uncertainty fraction is also 0.10 (10 percent) for ΔS because changes in surface-water storage are based only on approximate reservoir water-level altitude/volume relations rather than direct measurements. The largest estimated uncertainty fraction is 0.20 (20 percent) for E, which is based on differences between alternative methods for estimating evaporation both at Sand Hollow and other areas (Heilweil and others, 2007; Rosenberry and others, 2007).

The first two reports documenting monthly groundwater recharge beneath Sand Hollow Reservoir through August

2006 (Heilweil and others, 2005; Heilweil and Susong, 2007) did not include precipitation falling directly on the reservoir. Beginning with the third report (Heilweil and others, 2009), and continuing in this report, an additional term for precipitation falling directly on the reservoir (P) was included in equation 1. The monthly amount of precipitation falling on the reservoir is calculated by multiplying the total monthly precipitation recorded by the Sand Hollow weather station by the average reservoir surface area for that month (based on reservoir water-level altitude/area relations for the reservoir) (Washington County Water Conservancy District, written commun., 2006; RBG Engineering, written commun., 2002). The precipitation term in equation 1, however, does not account for precipitation runoff to the reservoir. Because of high evaporation rates and permeable surficial soils, precipitation events seldom produce runoff that reaches the lower part of Sand Hollow (L. Jessop, Washington County Water Conservancy District, oral commun., 2001), where the reservoir is situated.

Monthly water-budget values for Sand Hollow Reservoir are given in table 2. Values are generally monthly averages or totals, except for reservoir altitude and storage, which are shown for the last day of each month. Values for "Monthly evaporation rate," "Monthly evaporation," and "Monthly groundwater recharge" from March 2002 through January 2005 and from January 2008 through December 2009 are monthly averages; during February 2005 through December 2007, however, the values are the sum of daily measurements. Summing of daily evaporation estimates was discontinued after 2007 because comparison of daily and average monthly calculations during 2008 and 2009 showed little difference, and the equation used for calculating evapotranspiration is more appropriate for calculating average evaporation over longer time periods.

Changes in Reservoir Storage

Changes in reservoir storage were calculated from daily reservoir water-level altitude reported by the WCWCD using altitude/volume relations for the reservoir (RBG Engineering, written commun., 2002). Since inception of the reservoir in 2002, surface-water storage increased to a maximum of about 51,000 acre-ft in May of 2006. From the latter half of 2006 through 2007, surface-water storage decreased to about 32,000 acre-ft, and during 2008 and 2009, surface-water storage varied between about 31,000 and 44,000 acre-ft (table 2).

Reservoir Evaporation

The McGuinness and Bordne (1971) version of the Jensen-Haise method was selected for calculating evaporation from Sand Hollow Reservoir during this study. A detailed comparison to results using other methods for estimating evaporation is given in Heilweil and others (2005). The McGuinness and Bordne version of the Jensen-Haise method is based on the relation (McGuinness and Bordne, 1971):

$$PET = \{[((0.01T_a) - 0.37)(Q_s)]0.000673\}2.54 \qquad (3)$$

where

PET	is potential evaporation, in centimeters per day,
T_a	is air temperature, in degrees Fahrenheit, and
Q_s	is solar radiation, in calories per square centimeter per day.

The units for PET can be converted to feet per day by multiplying by 0.0328.

By using air temperature and solar radiation from the nearby weather station (fig. 2), monthly evaporation rates were calculated with equation 3. These estimated evaporation rates ranged from 0.04 to 0.97 ft per month from March 2002 through December 2009 (table 2; Heilweil and others, 2005; Heilweil and Susong, 2007; Heilweil and others, 2009). Multiplying the estimated evaporation rates by average reservoir surface area yields monthly evaporation losses that ranged from about 20 to 1,200 acre-ft between March 2002 and December 2009.

Estimates of Recharge from Sand Hollow Reservoir

Monthly estimates of evaporation (E), inflows (I_{sw}), outflows (O_{sw}), and changes in surface-water storage (ΔS) were used in equation (1) to calculate recharge to the Navajo Sandstone aquifer beneath Sand Hollow Reservoir. Monthly recharge from March 2002 through December 2009 ranged from about 50 to 3,500 acre-ft, with 2 standard deviation (σ) composite uncertainties ranging from about 6 to 14 percent of the estimate (table 2, fig. 9). Higher composite uncertainties in the summer reflect the larger, weighted importance of evaporation losses, which have the highest uncertainty. Several monthly recharge values differ from previously reported values in Heilweil and others (2005), Heilweil and Susong (2007), and Heilweil and others (2009) because of the inclusion of both "Monthly drain and spring return flow to reservoir" and "Monthly outflow to Sand Hollow Resort" (both through the 60-in. pipeline) in the current estimates. Monthly fluctuations in recharge may be partly caused by (1) changes in viscosity associated with varying reservoir water temperature, and (or) (2) changes in clogging caused by trapped gas bubble exsolution or dissolution, biofilm growth and decay, and silt accumulation or reduction. These factors and processes, along with the composite uncertainty of each monthly recharge estimate (6 to 14 percent), can contribute to the variability in estimated monthly recharge.

Estimated average monthly recharge rates beneath Sand Hollow Reservoir ranged from about 0.001 to 0.43 ft per day between March 2002 and December 2009 (fig. 10). Although the graph shows large monthly fluctuations, three general periods can be observed. Period 1 (March through June 2002) had very high initial rates and then a rapid decrease as the vadose zone of the Navajo Sandstone became saturated and a hydraulic connection between the reservoir and aquifer was

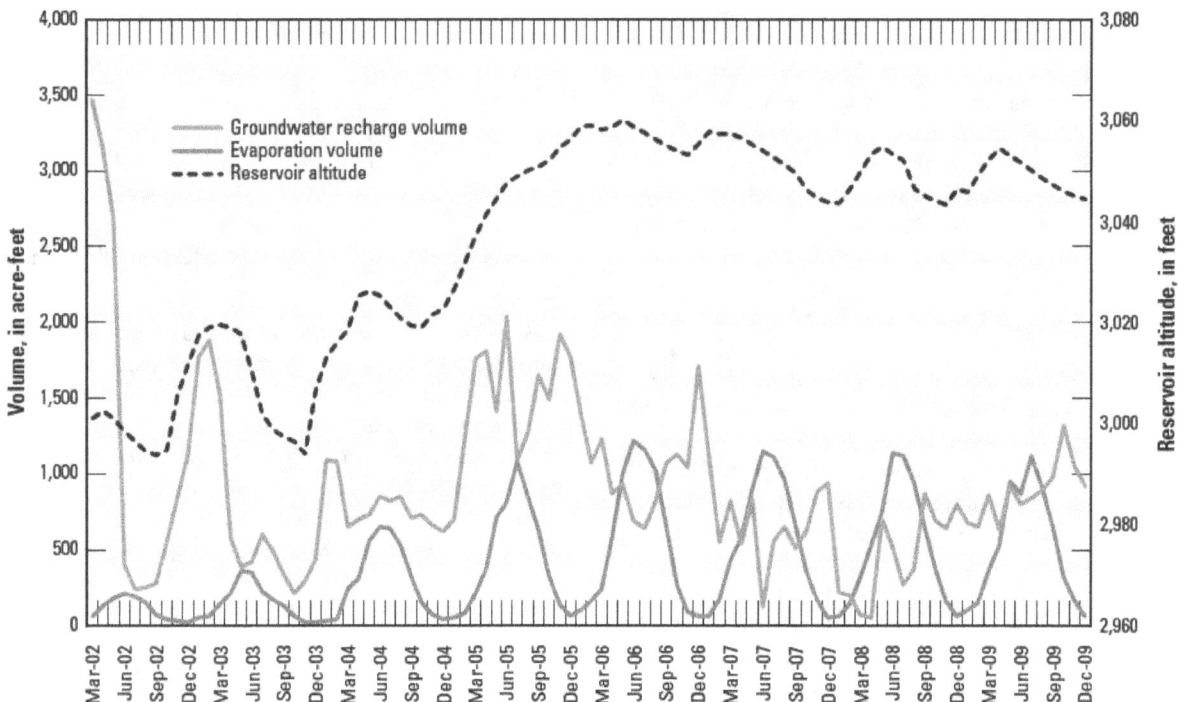

Figure 9.   Monthly estimated evaporation, recharge, and reservoir altitude, Sand Hollow Reservoir, Utah, 2002–09.

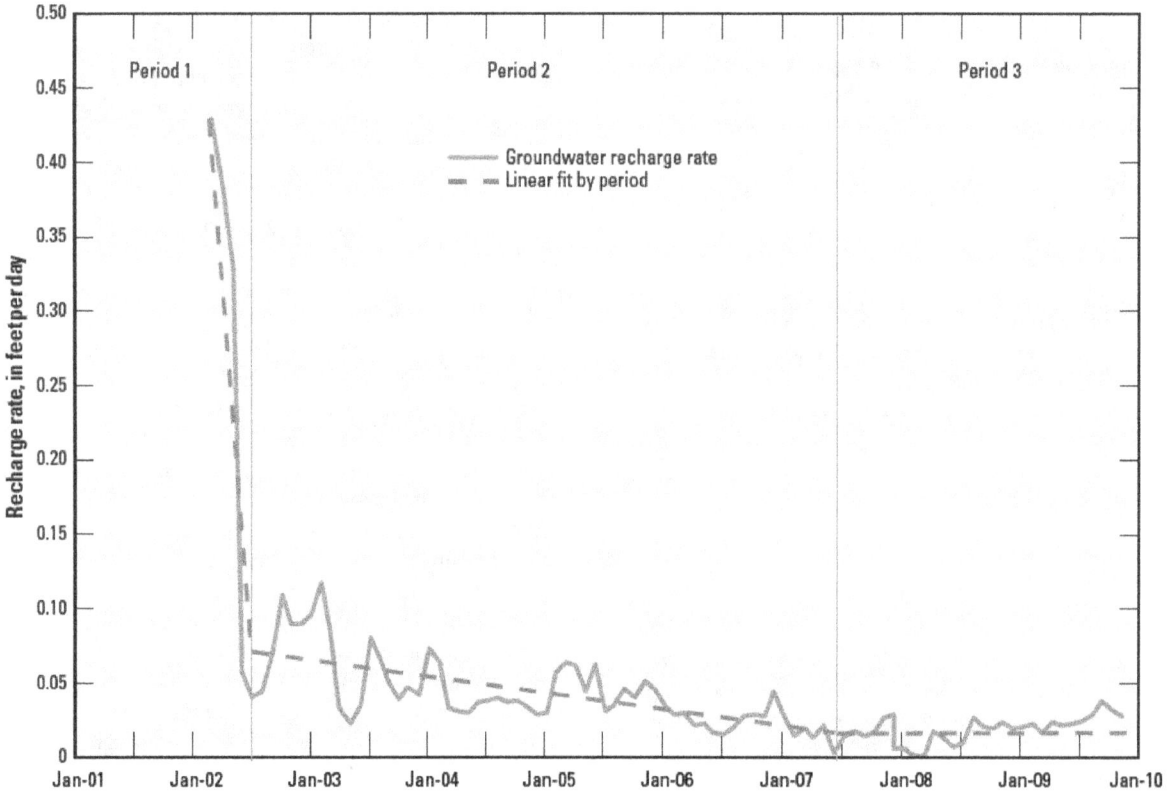

Figure 10.   Monthly calculated recharge rates from Sand Hollow Reservoir, Utah, 2002–09.

established, causing an abrupt decrease in hydraulic gradient. This establishment of a saturated hydraulic connection is supported by measurements in monitoring wells closest to the reservoir, which show rapidly rising water levels beginning in late spring 2002 near the southern end of the reservoir (fig. 5). Although consecutive monthly recharge rates occasionally fluctuate by more than 100 percent, Period 2 (mid-2002 through mid-2007) generally shows a gradual decline in recharge rates, while Period 3 (mid-2007 through 2009) demonstrates relatively constant recharge rates.

Net annual inflow, evaporation, and groundwater recharge from Sand Hollow Reservoir from 2002 through 2009 are shown in figure 11. Total net inflow during this period was about 154,000 acre-ft, with annual inflow during this period ranging from about 800 acre-ft in 2007 to 56,000 acre-ft in 2005. The general increase in reservoir water-level altitude and area from 2002 to 2007 resulted in a steady increase in the volume of annual evaporation from about 1,000 acre-ft in 2002 to about 6,600 acre-ft in 2006, which then leveled off from 2007 through 2009. Total estimated evaporative losses from 2002 through 2009 were about 37,000 acre-ft. Annual recharge ranged from a low of about 5,000 acre-ft in 2008 to a high of about 18,000 acre-ft in 2005. Total estimated recharge from 2002 through 2009 was about 86,000 acre-ft, with a 2σ uncertainty of 9,600 acre-ft.

Evaluation of the Movement of Managed Aquifer Recharge and Geochemical Mixing in the Navajo Sandstone

Water-quality data from Sand Hollow Reservoir and surrounding monitoring wells were used as chemical tracers to assess the movement of groundwater recharge from the reservoir through the Navajo Sandstone aquifer. These included both field water-quality parameters measured in-situ at monitoring wells and surface-water sites, as well as water samples collected for laboratory chemical analysis. Field water-quality parameters included water temperature, specific conductance, pH, dissolved oxygen, and total dissolved-gas pressure. Laboratory chemical analyses evaluated for tracing reservoir recharge included major ions, chloride to bromide ratios (Cl:Br), dissolved organic carbon, arsenic, tritium, chlorofluorocarbons (CFCs), and sulfur hexafluoride (SF_6).

Data Collection Methods

Field water-quality sampling methods were previously described in Heilweil and others (2005) and Heilweil and Susong (2007) and follow standard USGS water-quality sampling protocols (Wilde and Radtke, 1998). Field parameters were measured with a multi-parameter sonde placed at the bottom of each 2-in. monitoring well within the screened interval, and in the reservoir at water depths of approximately 2 ft. The

multi-parameter sonde was too large to enter the 1-in. monitoring wells (North Dam 3A, WD 1, WD 4, WD 5, WD RJ) and the seven temporary piezometers installed in the shallow sediments beneath the reservoir (fig. 2: P 1-2 through P 1-30). Consequently, field measurements from these wells were made on site with a flow-through chamber connected to the discharge from either a Waterra or peristaltic pump; no total dissolved-gas pressure measurements were made at these sites. Prior to sample collection from monitoring wells for laboratory chemical analyses, three casing volumes were purged from each well. After purging each well, water was pumped into samples bottles and filtered as necessary.

Field Water-Quality Parameters

Field parameters were measured to provide an on-site indication of both surface- and groundwater quality. Several of these parameters have also been useful for identifying the arrival of reservoir recharge at groundwater monitoring wells, including total dissolved-gas (TDG) pressure, dissolved oxygen (DO), and specific conductance. TDG pressure is a measurement of all the dissolved gas in a water sample and typically is dominated by the major components of air (nitrogen, oxygen), along with biologically generated gases such as carbon dioxide and methane. High TDG pressure and DO indicate dissolution of air bubbles in the sediments and underlying sandstone as groundwater levels rise during the initial filling of the reservoir. Groundwater under high hydrostatic pressure passing by these trapped air bubbles dissolves these gases. Specific conductance can be used as a proxy for dissolved-solids content. The specific conductance of water in Sand Hollow Reservoir water is generally higher than native groundwater, and changes can be used as an indicator of the arrival of reservoir recharge. The higher specific conductance of the reservoir water is attributed to the elevated dissolved-solids content of the Virgin River (Heilweil and others, 2005). The source of much of the discharge to the Virgin River is groundwater discharge within Zion National Park that has traveled through the gypsum-rich Carmel Formation (Cordova, 1981).

TDG pressures at the three 2-in. monitoring wells closest to the reservoir (WD 6, WD 9, WD 11) have shown the arrival and passage of peak values associated with reservoir recharge. TDG pressures at these wells increased from background values of 700 to 850 millimeters mercury (mm Hg) to values of 1,600 to more than 2,250 mm Hg, or about 2 to 3 times atmospherically equilibrated concentrations (fig. 12, table 3). The multi-parameter sonde used for TDG pressure measurements relies on a pressure transducer that cannot measure pressures greater than 2,250 mm Hg and is not within its linear calibration range above about 1,500 mm Hg. Measurements less than 1,500 have an error of less than 5 percent. Measured TDG pressure values at WD 9 (located 55 ft from the reservoir) exceeded 2,250 mm Hg during February and April 2005, indicating peak arrival occurred about 3 years after inception of the reservoir. TDG pressures at WD 9 have since declined; recent measurements during 2009 and 2010 of

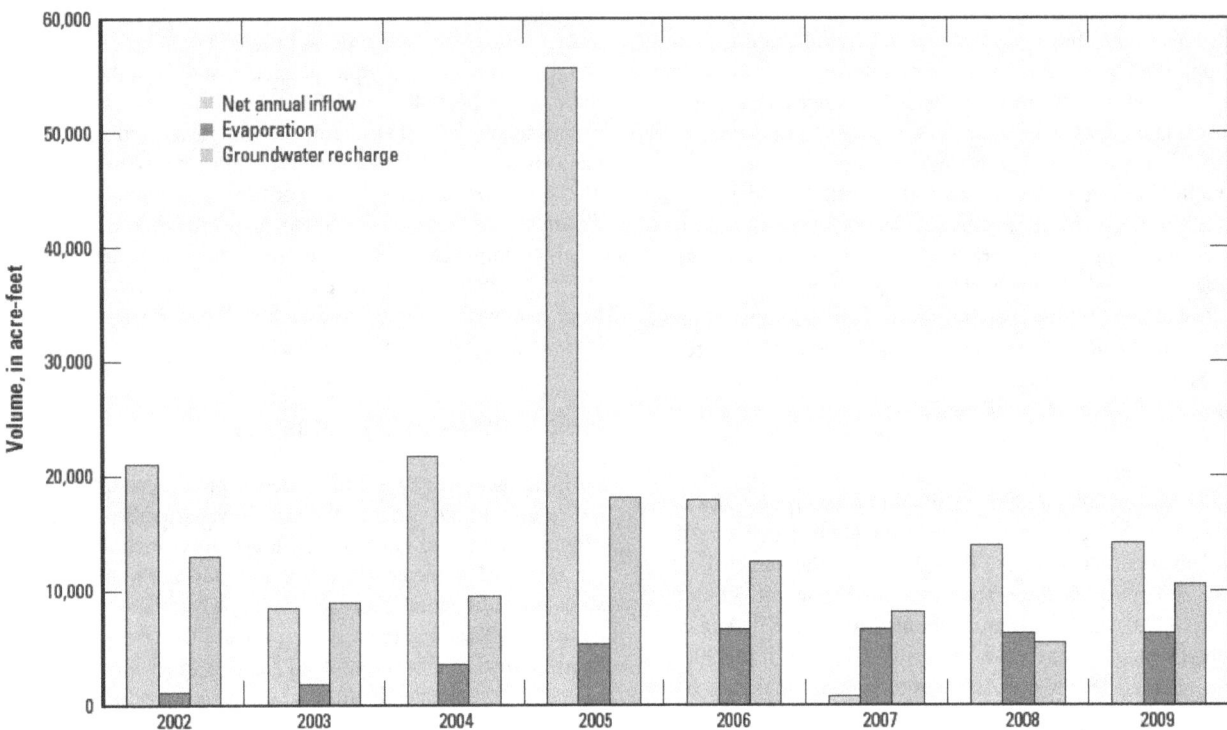

Figure 11.   Estimated net annual inflow, evaporation, and groundwater recharge from Sand Hollow Reservoir, Utah, 2002–09.

840 to 920 mm Hg are only slightly higher than the measured reservoir TDG and local barometric pressure of about 700 mm Hg. TDG pressures measured at WD 11 (located 160 ft from the reservoir) exceeded 2,250 mm Hg from 2005 through 2008, so an exact peak arrival date could not be determined. TDG pressures measured at WD 11 in 2010, although still elevated, declined to 1,650 mm Hg. TDG pressure measurements at WD 6 (located 1,000 ft from the reservoir) reached a peak value of about 1,800 mm Hg in April 2009 and declined to 1,200 mm Hg in 2010. Of the monitoring wells drilled in 2008, elevated TDG pressures have only been measured at WD 15 (1,300–1,500 mm Hg), located 2,400 ft from the reservoir. These elevated TDG pressures are likely caused by rising water levels and entrapment of air bubbles in the shallow part of the aquifer at this location rather than signifying the arrival of reservoir recharge.

Dissolved-oxygen concentrations at the three 2-in. monitoring wells closest to the reservoir (WD 6, WD 9, WD 11) show the arrival and passage of peak values associated with reservoir recharge (table 3). Background native groundwater DO values were generally between 6.1 and 8.7 mg/L. Similar to TDG pressure, DO reached elevated values, from about 18 to 25 mg/L (about 2 to 3 times atmospheric equilibration), in monitoring wells near the reservoir. DO at both WD 9 and WD 11 reached maximum values in April 2005; DO peaked in April 2009 at WD 6. While DO values have since declined at all three sites, WD 9 shows a much sharper decline, having values of less than 2 mg/L from 2008 to 2010. These values are much less than DO measurements of reservoir water,

which range from 7 to 12 mg/L, likely indicating biological consumption of oxygen in the shallow sediments beneath the reservoir.

Prior to inception of Sand Hollow Reservoir in March 2002, specific-conductance values of native groundwater ranged from 130 µS/cm at WD 6 to 560 µS/cm at WD RJ (table 3, fig. 13A). Elevated specific-conductance values at four monitoring wells (North Dam 3A, WD 6, WD 9, and WD 11) indicate the arrival of reservoir recharge. At WD 11 (located 160 ft from the reservoir), specific-conductance values reached a peak of 980 µS/cm in January 2006. This value is similar to the maximum measured value of water in the reservoir (1,000 µS/cm) and indicates about 4 years for the arrival of reservoir recharge. At WD 9 (located 55 ft from the reservoir), the peak measured value reached 1,230 µS/cm in January 2006; at WD 6 (located 1,000 ft from the reservoir), the peak measured value reached 1,330 µS/cm in June 2008. These peak specific-conductance values from WD 9 and WD 6, however, are higher than surface-water measurements in the reservoir and can indicate the mobilization of natural salts that accumulated in the vadose zone prior to the inception of the reservoir, rather than the arrival of reservoir recharge. These peaks in specific conductance either occurred as salt beneath the reservoir was flushed by reservoir recharge, or resulted from in-situ mobilization of salt near the monitoring wells when groundwater levels rose. Previous studies reported vadose-zone pore-water chloride values of up to 14,700 mg/L at borehole sites drilled in Sand Hollow prior to the construction of the reservoir (Heilweil and others, 2006). On the basis

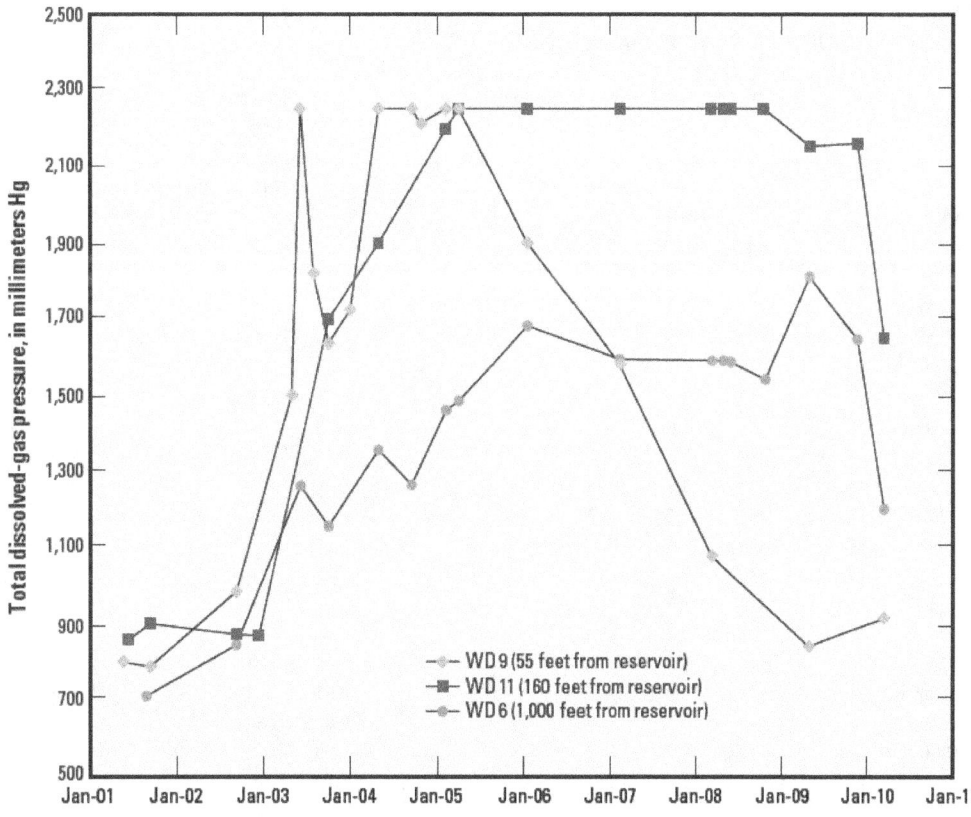

Figure 12.   Total dissolved-gas pressure in groundwater from selected monitoring wells in Sand Hollow, Utah.

Table 3.   Field water-quality parameters, dissolved organic carbon, tritium, chlorofluorocarbons, and sulfur hexafluoride in groundwater and surface water from Sand Hollow, Utah.

[Water temperature: C, degrees Celsius; Specific conductance: μS/cm, microsiemens per centimeter at 25 degrees Celsius; Dissolved oxygen, Dissolved organic carbon: mg/L, milligrams per liter; Total dissolved-gas pressure: mm Hg, millimeters mercury; Tritium, Tritium precision; TU, tritium units; CFC-11, CFC-12, CFC-113: pmol/kg, picomoles per kilogram; SF$_6$: fmol/kg, femtomoles per kilogram; %, percent; —, no data available; E, estimated; >, greater than; <, less than]

Site name	Date	Water temperature (°C)	Specific conductance (μS/cm)	pH (standard units)	Dissolved oxygen (mg/L)	Total dissolved-gas pressure (mm Hg)	[1]Dissolved organic carbon (mg/L)	[2]Tritium (TU)	[2]Tritium precision (TU)	[3]CFC-11 (pmol/kg)	[3]CFC-12 (pmol/kg)	[3]CFC-113 (pmol/kg)	[3]SF$_6$ (fmol/kg)	[2]Neon excess (%)
						Groundwater								
North Dam 3A	10/8/2002	15.9	4,430	8.0	5.0	—	—	2.71	0 14	—	—	—	—	—
	12/18/2002	14.7	2,830	8.0	10.8	—	—	—	—	—	—	—	—	—
	6/10/2003	21.5	1,330	7.8	—	—	—	—	—	—	—	—	—	—
	10/9/2003	—	1,230	7.8	—	—	—	—	—	—	—	—	—	—
	1/8/2004	16.0	1,220	8.2	—	—	—	—	—	—	—	—	—	—
	9/21/2004	18.4	978	7.7	11.0	—	—	—	—	—	—	—	—	—
	10/29/2004	15.9	905	7.9	11.1	—	—	—	—	—	—	—	—	—
	2/10/2005	15.3	961	7.7	13.5	—	—	—	—	—	—	—	—	—
	4/5/2005	16.5	964	7.8	12.6	—	—	—	—	—	—	—	—	—
	1/19/2006	—	835	8.0	—	—	—	—	—	—	—	—	—	—
	2/15/2007	15.2	840	7.9	7.5	—	—	2.53	0 31	—	—	—	—	—
	3/14/2008	14.8	820	7.7	4.0	—	—	3.45	0.44	—	—	—	—	—
	4/30/2009	—	850	7.2	—	—	—	3.03	0 11	—	—	—	—	—
	3/16/2010	22.8	864	7.6	1.3	—	1 91	3.05	0 12	0.54	2.0	0.10	—	—

Table 3. Field water-quality parameters, dissolved organic carbon, tritium, chlorofluorocarbons, and sulfur hexafluoride in groundwater and surface water from Sand Hollow, Utah.—Continued

[Water temperature: C, degrees Celsius; Specific conductance: µS/cm, microsiemens per centimeter at 25 degrees Celsius; Dissolved oxygen, Dissolved organic carbon: mg/L, milligrams per liter; Total dissolved-gas pressure: mm Hg, millimeters mercury; Tritium, Tritium precision: TU, tritium units; CFC-11, CFC-12, CFC-113: pmol/kg, picomoles per kilogram; SF_6: fmol/kg, femtomoles per kilogram; %, percent; —, no data available; E, estimated; >, greater than; <, less than]

Site name	Date	Water temperature (°C)	Specific conductance (µS/cm)	pH (standard units)	Dissolved oxygen (mg/L)	Total dissolved-gas pressure (mm Hg)	[1]Dissolved organic carbon (mg/L)	[2]Tritium (TU)	[2]Tritium precision (TU)	[3]CFC-11 (pmol/kg)	[3]CFC-12 (pmol/kg)	[3]CFC-113 (pmol/kg)	[3]SF_6 (fmol/kg)	[2]Neon excess (%)
WD 4	4/2/1999	21.0	355	8.2	—	—	—	0.22	0.10	—	—	—	—	—
	12/18/2002	18.7	350	7.7	8.1	—	—	—	—	—	—	—	—	—
	1/19/2006	—	345	8.0	—	—	—	—	—	—	—	—	—	—
	2/15/2007	19.0	340	7.9	8.7	—	—	—	—	—	—	—	—	—
	3/13/2008	22.6	350	7.8	7.8	—	—	0.25	0.10	—	—	—	—	—
	10/23/2008	21 2	360	8.0	—	—	—	0.13	0.10	0.62	0.61	0.09	0.44	—
	4/28/2009	—	350	7.8	—	—	—	0.15	0.07	0.54	0.52	0.07	0.45	—
	11/24/2009	18.7	338	7.8	9.5	—	0.434	0.09	0.03	0.42	0.54	0.07	—	—
	3/15/2010	19.7	362	7.7	9.5	—	E0.368	0.06	0.03	0.62	0.60	0.09	—	—
WD 5	4/3/1999	15.0	540	8.3	—	—	—	0.19	0.03	—	—	—	—	—
	12/17/2002	17.6	530	7.8	6.6	—	—	—	—	—	—	—	—	—
	1/18/2006	—	528	7.9	—	—	—	—	—	—	—	—	—	—
	2/15/2007	18 3	530	7.8	8.3	—	—	—	—	—	—	—	—	—
	3/13/2008	20.0	540	7.8	7.0	—	—	0.05	0.10	—	—	—	—	—
	10/23/2008	21.0	535	8.2	—	—	—	0.07	0.10	0.12	0.08	0.01	0.13	—
	4/30/2009	—	518	7.5	—	—	—	0.02	0.06	—	—	—	—	—
	11/24/2009	16 9	512	8.5	7.2	—	0.449	0.09	0.05	0.20	0.07	0.02	—	—
	3/15/2010	21.0	543	7.7	8.1	—	E0.438	0.09	0.10	0.19	0.10	0.03	—	—
WD RJ	4/2/1999	18.0	560	8.2	—	—	—	0.02	0.05	—	—	—	—	—
	12/17/2002	18 2	530	7.7	6.4	—	—	—	—	—	—	—	—	—
	1/18/2006	—	550	7.7	—	—	—	—	—	—	—	—	—	—
	2/15/2007	19.0	530	7.7	8.1	—	—	—	—	—	—	—	—	—
	3/12/2008	19 3	540	7.3	6.8	—	—	0.03	0.10	—	—	—	—	—
	4/28/2009	—	550	7.5	—	—	—	0.04	0.02	—	—	—	—	—
	3/15/2010	19.6	560	7.6	8.0	—	0.845	0.06	0.03	0.25	0.13	0.07	—	—
WD 6	5/15/2001	—	130	7.6	—	—	—	4.77	0.24	—	—	—	—	—
	8/28/2001	19.7	185	7.7	6.1	710	—	6.88	0.34	—	—	—	—	-0.2
	9/9/2002	19.4	290	7.7	—	850	—	—	—	—	—	—	—	13.5
	12/17/2002	19.0	400	7.6	9.3	920	—	—	—	—	—	—	—	—
	3/19/2003	19 2	424	7.5	10.9	1,150	—	—	—	—	—	—	—	—
	5/7/2003	19 3	450	7.5	—	1,220	—	—	—	—	—	—	—	—
	6/9/2003	19.6	390	7.8	14.0	1,260	—	—	—	—	—	—	—	—
WD 6	8/4/2003	19 3	350	7.5	11.9	1,280	—	—	—	—	—	—	—	—
	10/6/2003	19.6	400	7.6	12.0	1,160	—	—	—	—	—	—	—	—
	5/3/2004	19.4	697	7.4	15.2	1,357	—	—	—	—	—	—	—	—
	9/20/2004	19.6	824	7.7	15.0	1,266	—	—	—	—	—	—	—	75.8
	10/28/2004	19.0	810	7.6	13.5	1,240	—	—	—	—	—	—	—	—
	2/9/2005	19 2	447	7.9	14.6	1,460	—	—	—	—	—	—	—	83.6
	4/5/2005	19 2	462	7.6	15.5	1,490	—	—	—	—	—	—	—	88.0
	1/19/2006	18 9	684	7.6	17.7	[1]1,700	—	—	—	—	—	—	—	—
	2/15/2007	19 1	1,110	7.6	17.2	[1]1,600	—	—	—	—	—	—	—	—
	3/13/2008	19 2	1,300	7.5	14.4	[1]1,590	—	2.11	0.14	—	—	—	—	125.3
	4/29/2008	19 3	1,290	7.7	17.1	[1]1,590	—	—	—	—	—	—	—	124.0
	6/3/2008	19.4	1,330	7.6	16.5	[1]1,590	—	—	—	—	—	—	—	124.3
	10/24/2008	19.0	1,190	—	16.3	[1]1,540	—	2.55	0.13	2.8	1.3	0 15	0.72	—
	4/30/2009	19 2	1,040	7.7	22.0	[1]1,810	—	2.66	0.14	3.2	1.5	0 16	0.73	161.7
	11/23/2009	18 9	968	7.9	15.3	[1]1,650	1.71	2.93	0.23	1.7	1.8	0 17	—	140.8
	3/15/2010	19 2	923	7.5	14.4	1,200	1.68	3.15	0.15	1.7	1.6	0 19	—	88.5

Table 3. Field water-quality parameters, dissolved organic carbon, tritium, chlorofluorocarbons, and sulfur hexafluoride in groundwater and surface water from Sand Hollow, Utah.—Continued

[Water temperature: C, degrees Celsius; Specific conductance: µS/cm, microsiemens per centimeter at 25 degrees Celsius; Dissolved oxygen, Dissolved organic carbon: mg/L, milligrams per liter; Total dissolved-gas pressure: mm Hg, millimeters mercury; Tritium, Tritium precision: TU, tritium units; CFC-11, CFC-12, CFC-113: pmol/kg, picomoles per kilogram; SF₆: fmol/kg, femtomoles per kilogram; %, percent; —, no data available; E, estimated; >, greater than; <, less than]

Site name	Date	Water temperature (°C)	Specific conductance (µS/cm)	pH (standard units)	Dissolved oxygen (mg/L)	Total dissolved-gas pressure (mm Hg)	[1]Dissolved organic carbon (mg/L)	[2]Tritium (TU)	[2]Tritium precision (TU)	[3]CFC-11 (pmol/kg)	[3]CFC-12 (pmol/kg)	[3]CFC-113 (pmol/kg)	[3]SF₆ (fmol/kg)	[2]Neon excess (%)
WD 9	5/23/2001	19.5	295	7.7	8.0	800	—	0.00	0.01	—	—	—	—	—
	9/14/2001	19.4	280	7.4	—	790	—	0.20	0 15	—	—	—	—	49.5
	9/11/2002	19.5	345	7.9	—	980	—	—	—	—	—	—	—	28.1
	5/7/2003	19.7	315	7.8	—	[1]>2,250	—	—	—	—	—	—	—	—
	6/9/2003	19.5	350	7.7	24.4	[1]>2,250	—	—	—	—	—	—	—	158.8
	8/5/2003	19.7	720	7.5	19.3	[1]1,800	—	—	—	—	—	—	—	—
	10/7/2003	19.6	740	7.5	17.9	[1]1,600	—	—	—	—	—	—	—	—
	1/6/2004	19.4	630	7.7	16.7	[1]1,700	—	—	—	—	—	—	—	—
	5/3/2004	19.4	534	7.4	25.7	[1]>2,250	—	—	—	—	—	—	—	—
	9/20/2004	18.5	748	7.8	22.6	[1]>2,250	—	—	—	—	—	—	—	—
	10/28/2004	18.5	760	7.6	20.7	[1]2,210	—	—	—	—	—	—	—	—
	2/9/2005	18.4	779	7.7	20.2	[1]>2,250	—	—	—	—	—	—	—	246.2
	4/5/2005	18.5	815	7.4	23.2	[1]>2,250	—	—	—	—	—	—	—	—
	1/18/2006	18.0	1,230	7.9	15.0	[1]1,900	—	—	—	—	—	—	—	—
	2/14/2007	17.3	790	7.4	4.6	[1]1,600	—	—	—	—	—	—	—	—
	3/11/2008	17.0	816	7.3	1.5	1,080	—	2.61	0 22	—	—	—	—	138.4
	4/27/2009	16.6	832	7.4	1.8	840	—	2.99	0 12	1.2	2 2	0.19	2.15	91.4
	3/15/2010	16.4	842	7.3	1.7	920	1 23	3.20	0 14	0.8	2 2	0.21	—	103.4
(replicate)	3/15/2010	16.4	842	7.3	1.7	920	1 17	2.90	0 12	0.8	2 2	0.18	—	—
WD 11	6/14/2001	18.5	420	7.8	8.1	860	—	—	—	—	—	—	—	—
	9/14/2001	18.5	450	7.7	8.6	900	—	0.53	0.08	0.53	0 24	—	—	70.9
	9/12/2002	18.5	465	7.6	—	873	—	—	—	—	—	—	—	26.8
	12/16/2002	18.2	455	7.6	8.1	890	—	—	—	—	—	—	—	—
	5/7/2003	18.4	624	7.7	—	[1]1,770	—	—	—	—	—	—	—	—
	6/9/2003	18.4	650	7.9	22.5	[1]1,600	—	—	—	—	—	—	—	87.9
	8/5/2003	18.6	700	7.8	12.4	[1]1,520	—	—	—	—	—	—	—	—
	10/7/2003	18.5	800	7.8	19.4	[1]1,700	—	—	—	—	—	—	—	—
	5/3/2004	18.4	680	7.7	21.5	[1]1,900	—	—	—	—	—	—	—	—
	9/20/2004	18.0	922	8.2	23.5	[1]>2,250	—	—	—	—	—	—	—	—
	10/28/2004	18.0	993	7.9	22.8	[1]2,080	—	—	—	—	—	—	—	—
	2/9/2005	18.0	960	8.1	22.1	[1]2,200	—	—	—	—	—	—	—	162.5
	4/5/2005	17.8	929	7.9	25.2	[1]>2,250	—	—	—	—	—	—	—	—
	1/18/2006	17.6	977	7.9	23.0	[1]>2,250	—	—	—	—	—	—	—	—
	2/14/2007	17.1	820	7.6	19.0	[1]>2,250	—	—	—	—	—	—	—	—
	3/11/2008	17.0	840	7.6	14.9	[1]>2,250	—	2.30	0 14	—	—	—	—	319.1
	4/30/2008	17.0	840	7.7	17.4	[1]>2,250	—	—	—	—	—	—	—	—
	6/2/2008	17.1	850	7.7	18.9	[1]>2,250	—	—	—	—	—	—	—	213.8
	10/22/2008	16.7	836	8.0	15.9	[1]>2,250	—	2.36	0 11	—	—	—	—	—
	4/30/2009	15.9	843	7.7	19.4	[1]2,160	—	3.06	0 14	2.0	3.0	0.34	3.5	291.3
	11/23/2009	16.3	835	7.9	13.2	[1]2,160	1.46	2.75	0 12	0.8	3.0	0.30	—	293.7
	3/15/2010	16.2	837	7.7	10.3	[1]1,650	1 35	2.81	0 13	0.8	2 9	0.30	—	76.1
WD 15	10/25/2008	18.8	715	—	14.2	1,300	—	—	—	—	—	—	—	—
	4/28/2009	18.9	707	8.0	17.6	1,490	—	0.77	0.04	2.3	1 9	0.23	1.4	92.9
	11/23/2009	18.8	729	8.3	14.5	1,410	2.47	0.68	0.05	1.0	1 9	0.22	—	88.5
	3/16/2010	19.1	734	7.9	11.5	1,320	2.49	0.72	0.05	1.2	2 1	0.25	—	73.7
WD 16	10/25/2008	18.7	467	8.0	7.7	780	—	—	—	—	—	—	—	—
	4/27/2009	18.7	444	7.7	8.7	970	—	0.02	0.02	0.28	0 13	0.04	0.43	27.0
	11/24/2009	18.7	449	7.7	7.1	760	<0.66	0.03	0.04	0.12	0.01	0.01	—	10.6
	3/16/2010	18.7	441	7.6	5.1	770	<0.66	0.03	0.02	0.13	0.04	0.01	—	5.6

Table 3. Field water-quality parameters, dissolved organic carbon, tritium, chlorofluorocarbons, and sulfur hexafluoride in groundwater and surface water from Sand Hollow, Utah.—Continued

[Water temperature: C, degrees Celsius; Specific conductance: µS/cm, microsiemens per centimeter at 25 degrees Celsius; Dissolved oxygen, Dissolved organic carbon: mg/L, milligrams per liter; Total dissolved-gas pressure: mm Hg, millimeters mercury; Tritium, Tritium precision: TU, tritium units; CFC-11, CFC-12, CFC-113: pmol/kg, picomoles per kilogram; SF$_6$: fmol/kg, femtomoles per kilogram; %, percent; —, no data available; E, estimated; >, greater than; <, less than]

Site name	Date	Water temperature (°C)	Specific conductance (µS/cm)	pH (standard units)	Dissolved oxygen (mg/L)	Total dissolved-gas pressure (mm Hg)	[1]Dissolved organic carbon (mg/L)	[2]Tritium (TU)	[2]Tritium precision (TU)	[3]CFC-11 (pmol/kg)	[3]CFC-12 (pmol/kg)	[3]CFC-113 (pmol/kg)	[3]SF$_6$ (fmol/kg)	[2]Neon excess (%)
WD 18	4/28/2009	19.7	500	7.4	7.5	870		0.04	0.02	0.21	0.16	0.03	1.46	26.6
	3/16/2010	19 3	467	7.4	4.9	740	E0.48	0.14	0.02	0.16	0.07	0.01	—	36.9
WD 20	10/23/2008	19 1	341	—	7.9	740	—	—	—	—	—	—	—	—
	4/29/2009	19.7	331	7.5	6.7	760	—	0.06	0.01	0.10	0.02	0.01	0.01	-2.7
(replicate)	4/29/2009	19.7	331	7.5	6.7	760		0.06	0.01	0.12	0.02	0.01	0.06	—
	3/17/2010	19.4	344	7.4	7.2	720	<0.66	0.03	0.04	0.11	0.04	0.01	—	-2.4
Reservoir water														
Haul Road	9/10/2002	24 2	1,000	8.8	—	—	—	2.47	0.12	—	—	—	—	—
	12/18/2002	7 9	860	8.4	10.2	670	—	—	—	—	—	—	—	—
	3/20/2003	11 1	830	8.2	8.4	680	—	—	—	—	—	—	—	—
	6/10/2003	23.6	850	8.2	8.8	680	—	—	—	—	—	—	—	—
	8/6/2003	26.0	930	7.6	—	690	—	—	—	—	—	—	—	—
	10/7/2003	21 9	910	8.4	—	—	—	—	—	—	—	—	—	—
	1/8/2004	7 1	870	8.4	11.7	720	—	—	—	—	—	—	—	—
Boat Ramp	5/5/2004	17 3	710	8.2	8.5	680	—	—	—	—	—	—	—	—
	9/22/2004	18 9	766	8.5	7.2	—	—	—	—	—	—	—	—	—
	2/10/2005	8 3	857	8.4	11.3	—	—	—	—	—	—	—	—	—
	1/18/2006	6 9	815	8.5	11.9	—	—	—	—	—	—	—	—	—
	2/14/2007	5 1	760	8.1	11.6	—	—	—	—	—	—	—	—	—
	3/13/2008	9.6	820	8.4	10.1	—	—	—	—	—	—	—	—	—
	10/21/2008	18 3	819	8.7	8.9	700	—	3.59	0.18	2.3	1.5	0 22	1.49	—
	4/29/2009	16 1	790	8.4	7.0	—	—	4.61	0.20	3.1	2.0	0 32	1.94	—
	8/10/2009	25.0	800	8.6	—	—	2.85	—	—	—	—	—	—	—
	11/24/2009	11 3	797	8.5	9.5	—	2.95	3.29	0.14	2.1	2.6	0 30	—	—
	3/16/2010	9.8	817	8.0	9.4	—	2.88	3.64	0.15	3.0	3.3	0.47	—	—
SH1-18	10/23/2008	18.0	819	8.7	9.1	690	—	4.60	0.34	2.5	1.6	0 23	1.16	—
	4/29/2009	14 3	800	8.6	9.6	—	—	2.55	0.22	3.4	2.1	0 26	1.98	—
	8/10/2009	25 3	800	8.7	9.1	—	5.67	—	—	—	—	—	—	—
	3/16/2010	9.6	819	8.0	9.6	—	2.87	3.68	0.13	3.0	3.2	0.44	—	—
Shallow piezometers beneath reservoir														
P 1-14	4/29/2008	13.0	1,030	7.7	—	790	—	3.42	0.25	—	—	—	—	—
	10/21/2008	18 1	821	8.7	9.0	680	—	3.27	0.22	—	—	—	—	—
	8/10/2009	26.0	806	8.6	5.1	700	4.02	—	—	—	—	—	—	—
P 1-18	10/21/2008	18 1	818	8.6	9.1	690	—	3.58	0.18	—	—	—	—	—
	8/10/2009	25.4	803	8.6	4.3	690	4.24	—	—	—	—	—	—	—
P 1-25	10/21/2008	18.0	819	8.7	8.9	690	—	3.27	0.23	—	—	—	—	—
	8/10/2009	25 2	890	7.0	0.2	1,010	7.94	—	—	—	—	—	—	—
P 1-30	8/10/2009	24.8	950	8.6	—	—	13.52	—	—	—	—	—	—	—

[1]Dissolved organic carbon analyzed at the U S Geological Survey National Water Quality Laboratory, Denver, Colorado
[2]Tritium and neon analyzed at the University of Utah Dissolved Gas Laboratory, Salt Lake City, Utah
[3]CFC-11, CFC-12, CFC-113, and SF$_6$ analyzed at the U S Geological Survey Chlorofluorocarbon Laboratory, Reston, Virginia
[4]Total dissolved-gas pressures greater than 1,500 mm Hg exceed the linear calibration of the multi-parameter sonde

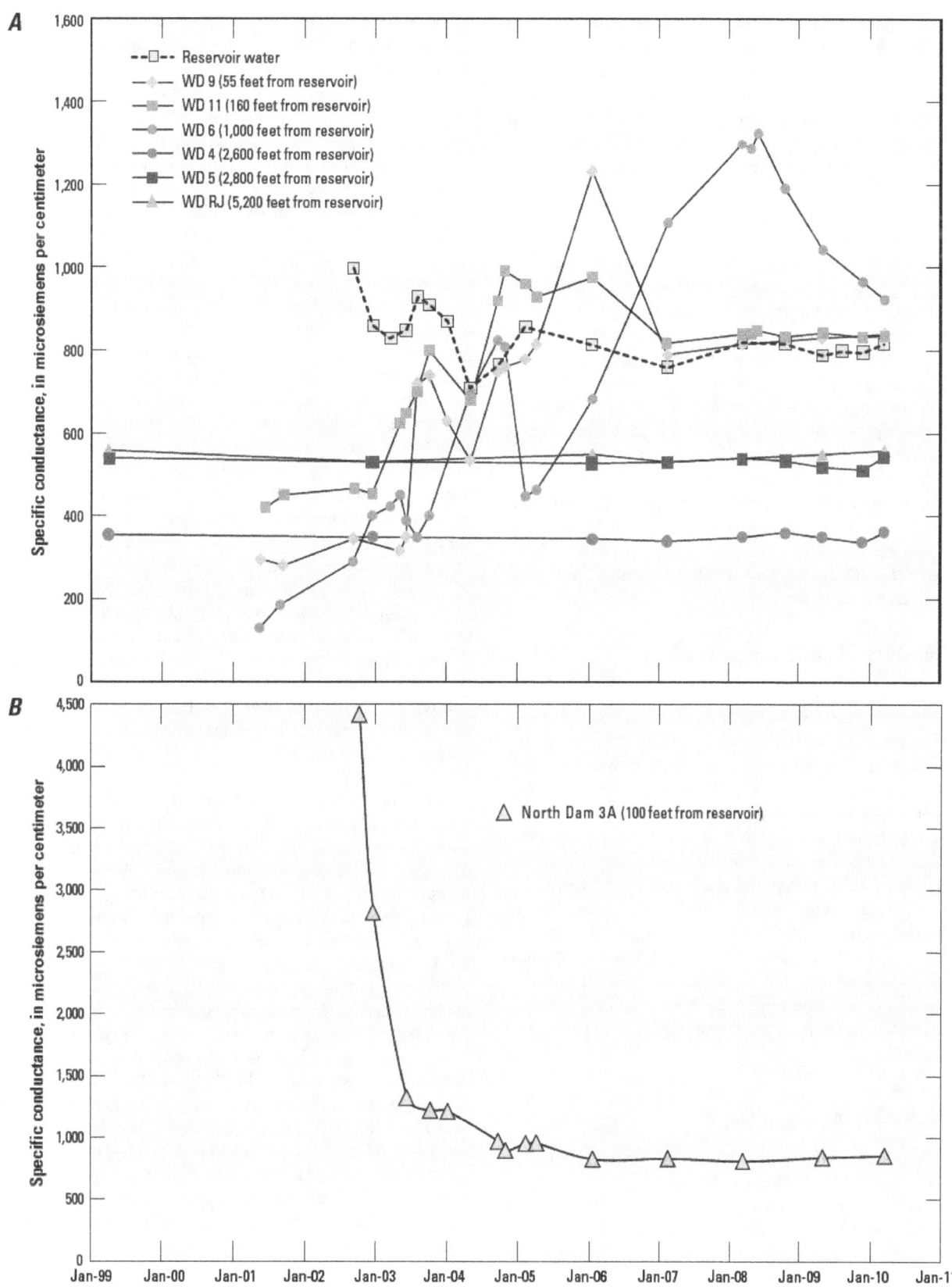

Figure 13. Specific conductance of *A*, reservoir water and groundwater from selected monitoring wells; and *B*, groundwater from the North Dam 3A monitoring well, Sand Hollow, Utah.

of a chloride to specific conductance (Cl:SpC) ratio of 0.27 from 76 vadose-zone pore-water samples in Sand Hollow (R_2 = 0.99; Heilweil and others, 2007), specific-conductance values of up to about 55,000 µS/cm are estimated. Flushing or mobilization of vadose-zone salt was observed at the North Dam 3A monitoring well, located 100 ft from the reservoir. This shallow well was only drilled to a total depth of 25.7 ft, which was similar to the shallow depths (up to 30 ft) where maximum vadose-zone salt accumulation was found (Heilweil and others, 2006). A peak specific-conductance value of 4,430 µS/cm was measured in the North Dam 3A well in October 2002 (fig. 13*B*), which was more than four times higher than any reservoir water measurements. After reaching peaks in 2006, specific-conductance values of water in both WD 9 and WD 11 declined to about 800 µS/cm, levels similar to values of the reservoir water (fig. 13*A*). In 2009 and 2010, specific-conductance values at WD 6 declined from the 2008 peak, but remained higher than reservoir water. WD 15, located about 2,500 ft from the reservoir, has shown a small increase in specific conductance since it was installed in 2008, likely from the mobilization of nearby vadose-zone salts rather than indicating the arrival of reservoir recharge. The other monitoring wells, farther from the reservoir (WD 4, WD 5, WD RJ, WD 16, WD 18, WD 20), have not shown a substantial increase in specific conductance through 2010.

Laboratory Chemical Analyses

Laboratory water-chemistry analyses of surface water from Sand Hollow Reservoir and groundwater from the Navajo Sandstone aquifer included major and minor dissolved inorganic ions, along with isotopes and other dissolved constituents that are potential tracers of reservoir recharge. The major inorganic ions included calcium, magnesium, sodium, potassium, bicarbonate, sulfate, chloride, and nitrate. Minor ions included fluoride, bromide, iron, manganese, arsenic, nitrite, ammonia, and orthophosphate. The isotopes and other dissolved constituents included dissolved organic carbon, tritium, deuterium, oxygen-18, and dissolved gases, such as chlorofluorocarbons (CFC–11, CFC–12, CFC–113), sulfur hexafluoride (SF_6), and noble gases (He, Ne, Ar, Kr, Xe). Since 2009, a set of replicates for all constituents has been separately analyzed yearly at one randomly selected sampling site for quality assurance.

Dissolved Inorganic Ions

Major-ion chemical signatures can be used to graphically evaluate the arrival of reservoir recharge. Figure 14*A* indicates that groundwater at some sites in Sand Hollow has been affected by recharge from the reservoir, while groundwater at other sites has not. The trilinear (Piper) diagram shows that native (background) groundwater in Sand Hollow is lower in chloride, magnesium, and sodium (plus potassium) and higher in calcium and bicarbonate than the infiltrating reservoir water. Both groundwater affected by reservoir recharge and pore

water from piezometers beneath the reservoir have chemical signatures that lie between the native groundwater and the reservoir water.

There are two sets of samples that do not plot in the "native groundwater" or "reservoir water" regions of figure 14*A*: (1) pore-water samples collected in June 2008 from temporary piezometers P 1-14 and P 1-18, installed in soils beneath the reservoir; and (2) samples collected in 2002 from the North Dam 3A monitoring well shortly after inception of the reservoir. While most of the shallow piezometer samples have major-ion chemical signatures similar to reservoir water, the June 2008 P 1-14 and P 1-18 temporary piezometer samples both have higher calcium and bicarbonate and lower sodium, chloride, and sulfate than the other shallow piezometer samples. This can indicate dissolution of native calcium carbonate, which accumulated as calcrete deposits at the soil/bedrock contact prior to construction of the reservoir (Heilweil and Solomon, 2004). The two samples collected from North Dam 3A had higher chloride and magnesium and lower calcium, sulfate, and bicarbonate than the reservoir and native groundwater samples. This is consistent with the very high specific-conductance values indicating that these samples were affected by mobilization of salts that had accumulated in the vadose zone prior to reservoir construction.

Apart from the samples collected in 2002 from North Dam 3A, major-ion chemical signatures can be used to identify the arrival of reservoir recharge from Sand Hollow Reservoir at downgradient monitoring wells. Figure 14*B* shows the evolution from native groundwater to water influenced by reservoir recharge at three monitoring wells close to the reservoir (WD 6, WD 9, and WD 11). Water samples from these wells have shifted from calcium carbonate- to sodium chloride-type water.

Chloride concentrations in Sand Hollow Reservoir water, ranging from 50 to 76 mg/L, are higher than in natural groundwater (0 to 48 mg/L; table 4). Like specific conductance, however, chloride concentrations may be problematic for interpreting the peak arrival of reservoir recharge. As discussed above, vadose-zone pore water from boreholes drilled prior to the reservoir had very high chloride concentrations, up to 14,700 mg/L (Heilweil and others, 2006).

Chloride to bromide ratios (Cl:Br) of water from Sand Hollow Reservoir, however, are more useful for tracing the movement of recharge from the reservoir through the aquifer than the use of chloride alone. Cl:Br ratios in the reservoir fluctuated between 1,100 and 5,000 from 2003 through 2006 (fig. 15). Beginning in 2007, the values have slowly increased from about 1,000 to 1,400. In contrast, Cl:Br ratios of native water in Sand Hollow are much lower: about 90 to 280 in groundwater (table 4) and about 125 to 250 in vadose-zone pore waters (Heilweil and others, 2006). Cl:Br ratios in wells receiving recharge from the reservoir (WD 9, North Dam 3A, WD 11, and WD 6) are between these two end members, and values have generally been rising in recent years, indicating the arrival of reservoir recharge. The highest groundwater Cl:Br ratio (about 1,060 ± 20) was in 2010 at WD 9, the closest monitoring well to the reservoir.

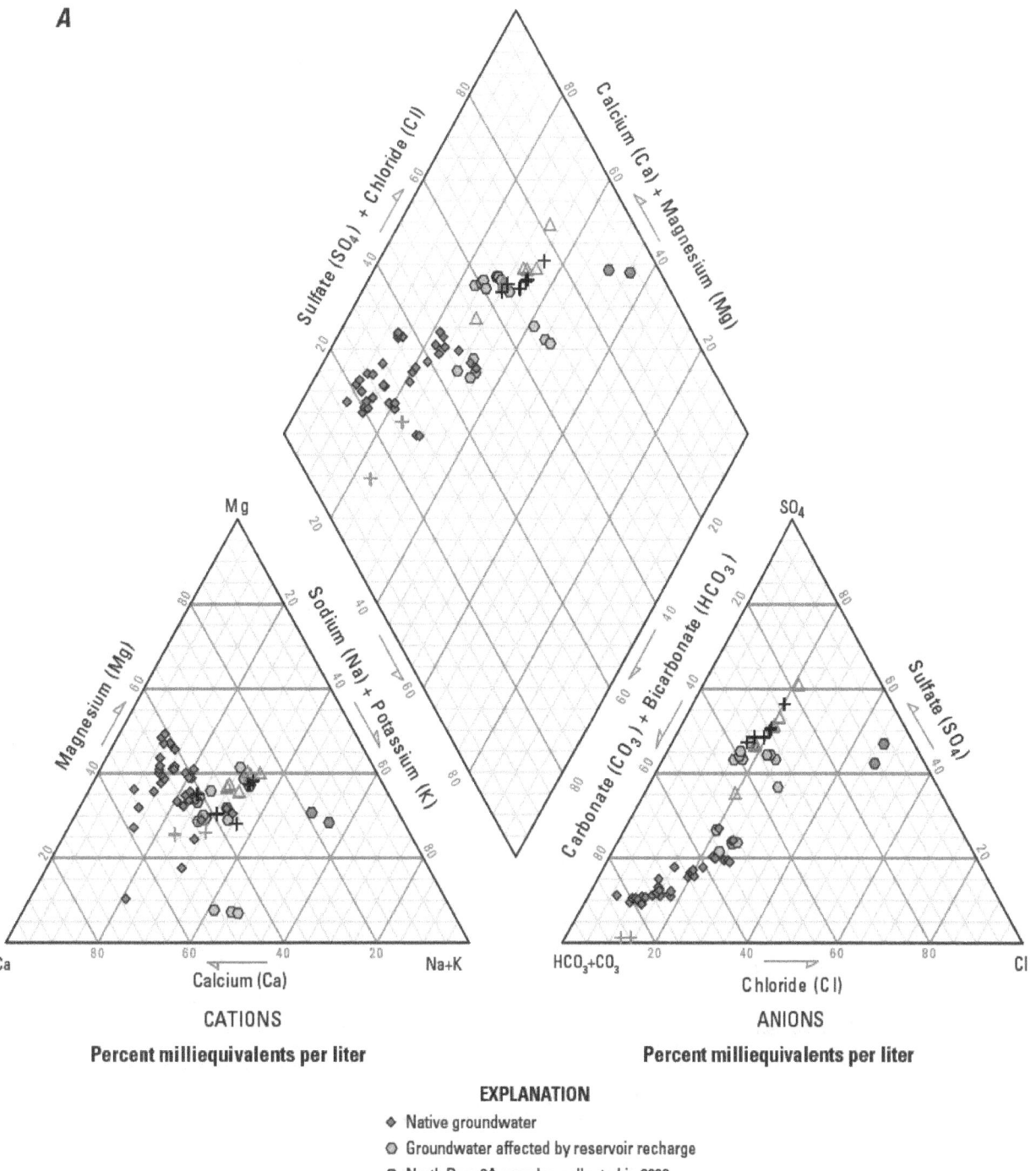

Figure 14. Major-ion chemistry from *A*, selected surface water and groundwater sites and *B*, monitoring wells near Sand Hollow Reservoir showing the evolution from native groundwater to recharge from the reservoir, Sand Hollow, Utah.

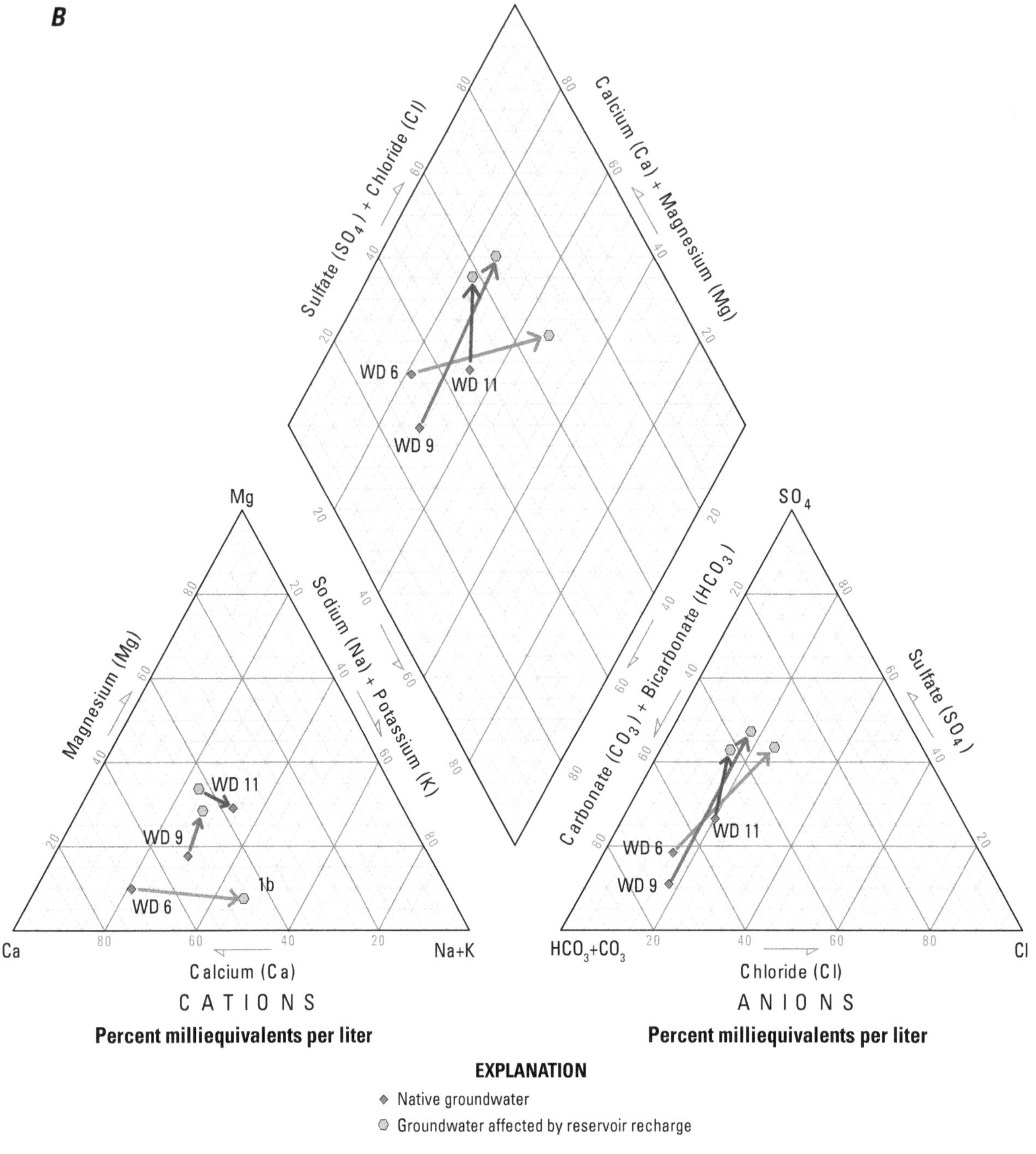

Figure 14. Major-ion chemistry from **B**, monitoring wells near Sand Hollow Reservoir showing the evolution from native groundwater to recharge from the reservoir, Sand Hollow, Utah.—Continued

Table 4. Major and minor chemical constituents in groundwater and surface water from selected sites in Sand Hollow, Utah.

[Specific conductance: µS/cm, microsiemens per centimeter at 25 degrees Celsius; mg/L, milligrams per liter; µg/L, micrograms per liter; Temperature: C, degrees Celsius; —, no data available; ft, feet; Cl:Br, chloride-to-bromide ratio; E, estimated; <, less than]

Site name	Date	Dissolved oxygen (mg/L)	Specific conductance (µS/cm)	pH	Temperature (°C)	Total dissolved solids (mg/L)	Calcium (mg/L as Ca)	Magnesium (mg/L as Mg)	Sodium (mg/L as Na)
				Native groundwater					
WD 4	12/18/02	8.1	350	7.7	18.7	205	29	17	16
	11/24/09	9.5	338	7.8	18.7	197	28.7	16.6	15.1
	3/15/10	9.5	362	7.7	19.7	217	27.3	16.2	13.3
WD 5	12/17/02	6.6	530	7.8	17.6	311	45	22	29
	11/24/09	7.2	512	8.5	16.9	298	43.1	21.4	27.2
	3/15/10	8.1	543	7.7	21.0	313	43.3	22.2	24.9
WD RJ	12/17/02	6.4	530	7.7	18.2	309	47	22	27
	3/15/10	8.0	560	7.6	19.6	338	46.6	22.6	25.1
WD 6	9/9/02	—	290	7.7	19.4	167	37	3.4	12
WD 7	9/10/01	9.8	380	7.8	18.8	—	37	12	25
WD 8	9/9/02	—	305	7.9	18.9	173	37	10	8 9
WD 9	9/11/02	—	335	7.9	19.5	189	36	7	22
WD 12	9/12/02	—	335	7.9	—	202	37	13	9.0
WD 13	8/30/01	—	275	8.1	19.9	—	24	16	8.4
WD 14	12/18/02	8.3	385	7.7	19.3	220	36	20	10
WD 15	4/28/09	17.6	707	8.0	18.9	414	41.0	35.9	48.0
	11/23/09	14.5	729	8.3	18.8	436	43.3	33.6	57.5
	3/16/10	11.5	734	7.9	19.1	458	42.0	33.8	51.6
WD 16	4/27/09	8.7	444	7.7	18.7	255	44.1	23.0	13.2
	11/24/09	7.1	449	7.7	18.7	260	42.3	21.9	13.7
	3/16/10	5.1	441	7.6	18.7	262	41.7	22.4	12 3
WD 18	4/28/09	7.5	500	7.4	19.7	280	45.2	19 5	24 5
	3/16/10	4.9	467	7.4	19.3	296	43.7	19 2	21 3
WD 20	4/29/09	6.7	331	7.5	19.7	188	30 2	17.4	11.7
(replicate)	4/29/09	6.7	331	7.5	19.7	188	30 3	18.0	11.7
	3/17/10	7.2	344	7.4	19.4	214	28.0	16.1	10.5
Well 1 at 890 ft	5/6/03	—	350	7.8	—	216	31	21	7.4
Well 2 at 400 ft	10/10/02	—	365	8.0	—	208	30	21	9.0
Well 2 at 615 ft	10/10/02	—	365	8.1	—	190	30	21	6.5
Well 2 at 750 ft	10/10/02	—	370	8.1	—	196	30	22	6.8
Well 4	8/29/01	—	480	8.0	20.1	—	36	19	38
	9/11/02	—	495	8.1	19.1	297	36	19	35
Well 8 at 245 ft	10/8/02	—	550	7.5	19.0	323	49	20	35
Well 9	8/30/01	—	285	7.9	20.7	179	27	16	7.0
[1]Hole N	5/25/01	—	310	7.8	—	—	34	14	4.4
[1]Slope 1a	9/12/01	8.7	240	7.9	19.5	—	26	13	8.3
	9/9/02	—	270	8.0	19.5	150	26	13	8 6
[1]Terracor 3	9/11/01	—	335	7.9	18.1	—	33	21	19
[1]Basin 2	8/27/01	9.4	290	7.8	18.9	—	30	13	8.5

Table 4. Major and minor chemical constituents in groundwater and surface water from selected sites in Sand Hollow, Utah.—Continued

[Specific conductance: μS/cm, microsiemens per centimeter at 25 degrees Celsius; mg/L, milligrams per liter; μg/L, micrograms per liter; Temperature: C, degrees Celsius; —, no data available; ft, feet; Cl:Br, chloride-to-bromide ratio; E, estimated; <, less than]

Site name	Date	Dissolved oxygen (mg/L)	Specific conductance (μS/cm)	pH	Temperature (°C)	Total dissolved solids (mg/L)	Calcium (mg/L as Ca)	Magnesium (mg/L as Mg)	Sodium (mg/L as Na)
Groundwater affected by reservoir recharge									
North Dam 3A	10/8/02	5.0	4,430	8.0	15.9	3,020	150	160	590
	12/18/02	10.8	2,830	8.0	14.7	1,890	110	110	340
	3/16/10	1.3	864	7.6	22.8	554	65.8	38.0	51.2
WD 6	4/30/09	22.0	1,040	7.7	19 2	660	98 5	9.0	113
	11/23/09	15.3	968	7 9	18 9	629	93.6	8.7	101
WD 6	3/15/10	14.4	923	7 5	19 2	618	94.1	8.6	86.3
WD 9	4/27/09	1.8	832	7.4	16.6	549	78.2	30.9	53.7
	3/15/10	1.7	842	7 3	16.4	543	71.8	31.0	52.2
(replicate)	3/15/10	1.7	842	7.3	16.4	545	67.4	28.8	51.0
WD 11	5/3/04	21.5	677	7.7	18.4	440	69.0	31.6	68.1
	4/30/09	19.4	843	7.7	15.9	557	79.2	38.6	49.6
	11/23/09	13 2	835	7.9	16.3	553	74.0	35.7	49.4
	3/15/10	10.3	837	7.7	16 2	552	67 2	34 3	45.6
Well 9	9/11/02	—	740	8 2	19 5	458	53	28	52
Sand Hollow Reservoir water									
Haul Road	9/10/02	—	1,000	8.8	24 2	669	63	43	71
Boat Ramp	5/5/04	8.5	710	8 2	17 3	442	63	26	45
	4/29/09	7.0	790	8.4	16 1	503	54.3	37.4	53.7
	11/24/09	9.5	797	8 5	11 3	502	40.9	39.8	62.9
	3/16/10	9.4	817	8.0	9.8	534	43.5	38.4	57.6
SH 1-18	4/29/09	9.6	800	8.6	14.3	502	56.1	37.2	53.6
	8/10/09	9.1	800	8.7	25.3	501	42.6	38.3	60.5
	3/16/10	9.6	819	8.0	9.6	525	45.9	40.8	58.6
Shallow piezometers beneath reservoir									
P 1-2	6/3/08	—	775	6.4	20.0	490	50.6	23.9	52.9
P 1-6	6/2/08	—	910	6.8	19.5	624	81.0	29.2	59.8
P 1-10	6/2/08	—	891	7.2	19.0	581	69.2	32.6	59.3
P 1-14	6/2/08	—	943	7.5	19.4	591	103 3	31.7	45 1
	8/10/09	5.1	806	8.6	26.0	497	46.7	36.8	61.1
P 1-18	6/2/08	—	828	7 3	19.0	517	75.0	39.0	48.1
	8/10/09	4.3	803	8.6	25.4	501	47.0	36.9	60.3
P 1-25	8/10/09	0.2	890	7.0	25 2	505	45.0	37.7	61.5
P 1-30	8/10/09	—	950	8.6	24.8	498	43.7	37.8	61.1

[1]Abandoned well beneath reservoir; see Heilweil and others (2005) for location and well information.

Table 4. Major and minor chemical constituents in groundwater and surface water from selected sites in Sand Hollow, Utah.—Continued

[Specific conductance: μS/cm, microsiemens per centimeter at 25 degrees Celsius; mg/L, milligrams per liter; μg/L, micrograms per liter; Temperature: C, degrees Celsius; —, no data available; ft, feet; Cl:Br, chloride-to-bromide ratio; E, estimated; <, less than]

Site name	Date	Potassium (mg/L as K)	Alkalinity as CaCO₃ (mg/L)	Sulfate (mg/L as SO₄)	Chloride (mg/L as Cl)	Fluoride (mg/L as F)	Bromide (mg/L as Br)	Cl:Br	Silica (mg/L as SiO₂)	Iron (μg/L as Fe)
					Native groundwater					
WD 4	12/18/02	2 1	125	18.1	18.8	0.23	0.08	235	14	<10
	11/24/09	2 1	121	20.6	17.2	—	0.10	179	14 3	E3.8
	3/15/10	2 1	129	19.7	17.9	0.25	0.10	184	15.7	<6
WD 5	12/17/02	1.8	138	46.8	44.8	0.29	0.16	280	13	<10
	11/24/09	1.8	136	46.4	37.9	0.273	0.23	168	13.4	<6
	3/15/10	2.0	136	45.8	39.2	0.279	0.24	164	15	<6
WD RJ	12/17/02	2 3	137	46	47.8	0.51	0.20	239	14	<10
	3/15/10	2 3	139	47.9	47.2	0.509	0.27	176	15 3	<6
WD 6	9/9/02	1.6	93	24	15.0	E0.08	0.16	94	13	<10
WD 7	9/10/01	1 9	137	28	18.0	0.3	0.13	139	14	<10
WD 8	9/9/02	2 3	116	15	10.1	0.1	0.07	144	14	<10
WD 9	9/11/02	1.6	120	18	21.4	0.5	0.06	357	15	9
WD 12	9/12/02	1.6	115	19	20.0	0.2	0.08	250	15	<10
WD 13	8/30/01	1 5	109	12	12.1	E0.1	0.05	258	12	<10
WD 14	12/18/02	2.4	122	29	28.3	0.25	0.11	257	13	<10
WD 15	4/28/09	2 1	191	71.4	57.0	0.41	0.33	174	15	<4
	11/23/09	2 1	184	80.4	63.5	0.41	0.36	178	14	<6
	3/16/10	2 1	188	84.7	68.8	0.42	0.36	189	15	<6
WD 16	4/27/09	1.9	136	33.6	29.1	0.251	0.17	170	14	<4
	11/24/09	1.7	129	33.8	28.7	0.214	0.18	158	13	<6
	3/16/10	1.8	135	33.0	29.9	0.224	0.18	169	13.8	<6
WD 18	4/28/09	1.9	143	40.4	36.1	0.367	0.21	171	16	16
	3/16/10	1.8	155	37.9	34.1	0.333	0.22	154	15.8	<6
WD 20	4/29/09	2 1	120	20.8	16.4	0.278	0.09	178	14	53
(replicate)	4/29/09	2 1	121	20.9	16.6	0.271	0.10	170	14	38
	3/17/10	1 9	120	19.6	17.2	0.244	0.09	185	14 2	<6
Well 1 at 890 ft	5/6/03	2 9	130	19	16.9	1.08	—	—	11	11
Well 2 at 400 ft	10/10/02	2 1	129	20	17.8	0.2	—	—	11	10
Well 2 at 615 ft	10/10/02	2 5	131	16	13.2	0.23	—	—	11	27
Well 2 at 750 ft	10/10/02	2.7	134	18	14.3	0.23	0.10	143	12	19
Well 4	8/29/01	2.0	128	58	44.4	E0.1	0.20	218	13	<10
	9/11/02	2.0	124	56	42.0	0.2	0.17	247	13	<10
Well 8 at 245 ft	10/8/02	2 1	141	70	38.7	0.29	0.15	258	14	<10
Well 9	8/30/01	1 9	115	13	13.0	0.2	0.07	186	13	<10
[1]Hole N	5/25/01	4.7	125	16	6.5	0.7	0.05	130	18	13
[1]Slope 1a	9/12/01	1.7	109	15	12.6	E0.1	0.05	277	14	<10
[1]Slope 1a	9/9/02	1.8	107	14	10.1	0.1	0.05	202	14	<10
[1]Terracor 3	9/11/01	1.6	136	31	28.7	0.3	0.14	203	13	<10
[1]Basin 2	8/27/01	2.4	115	13	9.9	E0.1	0.05	198	14	<10

Table 4. Major and minor chemical constituents in groundwater and surface water from selected sites in Sand Hollow, Utah.— Continued

[Specific conductance: µS/cm, microsiemens per centimeter at 25 degrees Celsius; mg/L, milligrams per liter; µg/L, micrograms per liter; Temperature: C, degrees Celsius; —, no data available; ft, feet; Cl:Br, chloride-to-bromide ratio; E, estimated; <, less than]

Site name	Date	Potassium (mg/L as K)	Alkalinity as CaCO₃ (mg/L)	Sulfate (mg/L as SO₄)	Chloride (mg/L as Cl)	Fluoride (mg/L as F)	Bromide (mg/L as Br)	Cl:Br	Silica (mg/L as SiO₂)	Iron (µg/L as Fe)
Groundwater affected by reservoir recharge										
North Dam 3A	10/8/02	2.0	148	1,020	744	0.9	41.2	18	13	<30
	12/18/02	3.6	155	584	476	0.79	2.44	195	14	<30
	3/16/10	3.0	177	187	55.2	0.43	0.06	882	16	<6
WD 6	4/30/09	1.6	169	220	92.5	0.32	0.31	295	13	<4
	11/23/09	1.5	161	210	80.3	0.30	0.28	286	12	<6
	3/15/10	1.5	166	211	77.9	0.32	0.24	322	13	20.8
WD 9	4/27/09	3.4	157	200	53.4	0.27	0.06	900	12	5
	3/15/10	3.3	157	200	56.6	0.24	0.05	1,040	12	13.6
(replicate)	3/15/10	3.2	155	198	56.3	0.27	0.05	1,083	12	13 5
WD 11	5/3/04	1.7	187	89.7	49.8	0.4	0.25	199	15	<6
	4/30/09	2.4	186	187	49.6	0.35	0.07	687	14	<4
	11/23/09	2.2	171	191	49.8	0.31	0.07	711	13	<6
	3/15/10	2.2	178	190	51.8	0.32	0.07	781	14	18.6
Well 9	9/11/02	2.3	124	126	72.2	0.2	0.28	258	14	250
Sand Hollow Reservoir water										
Haul Road	9/10/02	5.3	92	300	76.0	0.3	0.02	3,800	4.9	<10
Boat ramp	5/5/04	3.3	161	122	50.0	0.21	0.01	5,000	7.3	<6
	4/29/09	4.0	147	189	54.9	0.31	0.04	1,227	2.9	<4
	11/24/09	4.3	108	212	60.4	0.28	0.05	1,313	1.5	<6
	3/16/10	4.6	120	211	61.7	0.30	0.04	1,374	1.4	6 3
SH 1-18	4/29/09	4.2	146	190	54.6	0.27	0.04	1,318	3.0	<4
	8/10/09	4.3	110	—	—	0.24	—	—	1.3	3
	3/16/10	4.7	124	211	61.6	0.30	0.04	1,417	1.2	6 2
Shallow piezometers beneath reservoir										
P 1-2	6/3/08	7.8	85	195	51.5	0.07	0.05	1,120	7	—
P 1-6	6/2/08	7.6	175	218	55.4	0.10	0.08	735	9.9	—
P 1-10	6/2/08	4.5	159	217	56.9	0.14	0.06	922	11.7	—
P 1-14	6/2/08	18.4	458	6.3	44.7	0.10	0.17	262	13.2	—
	8/10/09	4.7	121	195	57.2	0.26	0.09	622	2.4	<4
P 1-18	6/2/08	3.8	343	52.2	50.1	0.19	0.11	446	14.7	—
	8/10/09	4.5	133	193	57.1	0.26	0.06	1,016	2.6	5.7
P 1-25	8/10/09	4.5	120	199	57.7	0.25	0.05	1,145	1.7	4 1
P 1-30	8/10/09	4.6	114	203	—	—	—	—	1.5	6.8

[1]Abandoned well beneath reservoir; see Heilweil and others (2005) for location and well information.

Table 4. Major and minor chemical constituents in groundwater and surface water from selected sites in Sand Hollow, Utah.—Continued

[Specific conductance: µS/cm, microsiemens per centimeter at 25 degrees Celsius; mg/L, milligrams per liter; µg/L, micrograms per liter; Temperature: C, degrees Celsius; —, no data available; ft, feet; Cl:Br, chloride-to-bromide ratio; E, estimated; <, less than]

Site Name	Date	Manganese (µg/L as Mn)	Arsenic (µg/L as As)	Nitrogen (nitrite + nitrate) (mg/L as N)	Nitrogen, nitrite (mg/L as N)	Nitrogen, ammonia (mg/L as N)	Phosphorus (ortho-phosphate) (mg/L as P)
				Native groundwater			
WD 4	12/18/02	<2	13.2	2.35	<0.008	<0.04	0.02
	11/24/09	<0.2	14.7	2.29	<0.002	<0.02	0.0266
	3/15/10	<0.2	14.4	2.29	<0.002	<0.02	0.0276
WD 5	12/17/02	E1	9.1	4.18	<0.008	<0.04	E0.01
	11/24/09	<0.2	9.6	4.61	<0.002	<0.02	0.0226
	3/15/10	<0.2	9.0	4.60	<0.002	<0.02	0.01
WD RJ	12/17/02	<2	7.9	3.28	<0.008	<0.04	0.01
	3/15/10	<0.2	8.3	3.28	<0.002	E0.012	0.0167
WD 6	9/9/02	E2	2.0	E1.6	<0.008	<0.04	0.02
WD 7	9/10/01	<3	6.0	3.80	<0.008	<0.04	0.02
WD 8	9/9/02	<2	6.0	3.90	<0.008	<0.04	0.02
WD 9	9/11/02	15	12.0	0.48	<0.008	<0.04	0.01
WD 12	9/12/02	1	10.0	2.10	<0.008	<0.04	0.02
WD 13	8/30/01	2	6.3	2.00	<0.006	<0.04	0.02
WD 14	12/18/02	<2	15.6	2.18	<0.008	<0.04	0.02
WD 15	4/28/09	0.7	28.3	3.32	E0.001	<0.02	0.02
	11/23/09	0.1	28.9	3.46	<0.002	<0.02	0.02
	3/16/10	<0.2	27.5	3.54	<0.002	<0.02	0.02
WD 16	4/27/09	<0.2	6.2	4.48	E0.001	<0.02	0.01
	11/24/09	<0.2	6.1	4.50	<0.002	<0.02	0.01
	3/16/10	0.79	5.9	4.44	<0.002	<0.02	0.010
WD 18	4/28/09	1	10.6	3.15	0.002	<0.02	0.01
	3/16/10	<0.2	10.0	3.14	<0.002	<0.02	0.016
WD 20	4/29/09	1	7.7	2.41	E0.001	<0.02	0.02
(replicate)	4/29/09	0.39	8.0	2.41	E0.001	<0.02	0.02
	3/17/10	0.53	8.0	2.40	<0.002	<0.02	0.018
Well 1 at 890 ft	5/6/03	19	9.1	3.37	0.008	0.03	0.01
Well 2 at 400 ft	10/10/02	12	2.6	3.41	0.008	0.10	0.02
Well 2 at 615 ft	10/10/02	6	4.6	3.73	0.004	<0.04	0.02
Well 2 at 750 ft	10/10/02	3	5.9	3.84	<0.008	0.03	0.02
Well 4	8/29/01	<3	7.1	1.50	<0.006	<0.04	0.02
	9/11/02	<2	8.0	2.10	<0.008	<0.04	0.02
Well 8 at 245 ft	10/8/02	5	16.6	1.72	0.03	0.18	0.01
Well 9	8/30/01	<3	12.4	2.40	<0.006	<0.04	0.02
[1]Hole N	5/25/01	6	18.1	0.67	0.01	0.03	0.02
[1]Slope 1a	9/12/01	<3	9.3	1.70	<0.006	<0.04	0.02
	9/9/02	1	10.2	1.70	<0.008	<0.04	0.02
[1]Terracor 3	9/11/01	2	12.8	2.70	<0.006	<0.04	0.02
[1]Basin 2	8/27/01	20	6.3	2.90	<0.006	<0.04	0.02

Table 4.   Major and minor chemical constituents in groundwater and surface water from selected sites in Sand Hollow, Utah.—Continued

[Specific conductance: µS/cm, microsiemens per centimeter at 25 degrees Celsius; mg/L, milligrams per liter; µg/L, micrograms per liter; Temperature: C, degrees Celsius; —, no data available; ft, feet; Cl:Br, chloride-to-bromide ratio; E, estimated; <, less than]

Site Name	Date	Manganese (µg/L as Mn)	Arsenic (µg/L as As)	Nitrogen (nitrite + nitrate) (mg/L as N)	Nitrogen, nitrite (mg/L as N)	Nitrogen, ammonia (mg/L as N)	Phosphorus (ortho-phosphate) (mg/L as P)
Groundwater affected by reservoir recharge							
North Dam 3A	10/8/02	<5	90.1	17.80	<0.008	0.03	0.03
	12/18/02	<5	63.9	14.30	<0.008	<0.04	0.03
	3/16/10	3.19	35.2	<0.04	E0.001	E0.01	0.03
WD 6	4/30/09	0.23	3.3	1.2	<0.002	<0.02	0.01
	11/23/09	<0.2	3.3	1.06	<0.002	<0.02	0.01
	3/15/10	0.32	3.0	0.97	<0.002	<0.02	0.01
WD 9	4/27/09	4	5.8	0.09	<0.002	<0.02	0.01
	3/15/10	0.68	6.1	0.09	<0.002	<0.02	0.01
(replicate)	3/15/10	0.57	6.1	0.09	<0.002	<0.02	0.01
WD 11	5/3/04	<0.8	15.3	3.06	<0.008	<0.04	0.02
	4/30/09	<0.2	9.6	0.99	<0.002	<0.02	0.01
	11/23/09	<0.2	10.3	0.67	<0.002	<0.02	0.01
	3/15/10	0.26	10.0	0.70	<.002	<0.02	0.02
Well 9	9/11/02	6	17.0	2.20	<0.008	<0.04	0.02
Sand Hollow Reservoir water							
Haul Road	9/10/02	<2	2.0	0.04	<0.008	<0.04	0.02
Boat Ramp	5/5/04	1.3	1.1	—	—	—	—
	4/29/09	0.3	1.4	0.04	0.002	<0.02	0.008
	11/24/09	0.2	1.6	<0.04	<0.002	<0.02	0.008
	3/16/10	1.7	1.3	E0.033	<0.002	0.02	<0.008
SH 1-18	4/29/09	0.4	1.4	0.04	0.003	0.13	0.008
	8/10/09	0.3	1.6	<0.04	<0.002	<0.02	0.008
	3/16/10	1.8	1.4	E0.025	<0.002	0.025	<0.008
Shallow piezometers beneath reservoir							
P 1-2	6/3/08	—	15.5	—	—	—	—
P 1-6	6/2/08	—	12.4	—	—	—	—
P 1-10	6/2/08	—	8.5	—	—	—	—
P 1-14	6/2/08	—	10.0	—	—	—	—
	8/10/09	43.88	2.3	<0.04	<0.002	<0.02	0.006
P 1-18	6/2/08	—	12.6	—	—	—	—
	8/10/09	52.77	2.7	<0.04	<0.002	<0.02	0.01
P 1-25	8/10/09	44.65	2.1	<0.04	<0.002	0.043	0.01
P 1-30	8/10/09	7.10	1.7	0.058	<0.002	<0.02	0.01

[1]Abandoned well beneath reservoir; see Heilweil and others (2005) for location and well information.

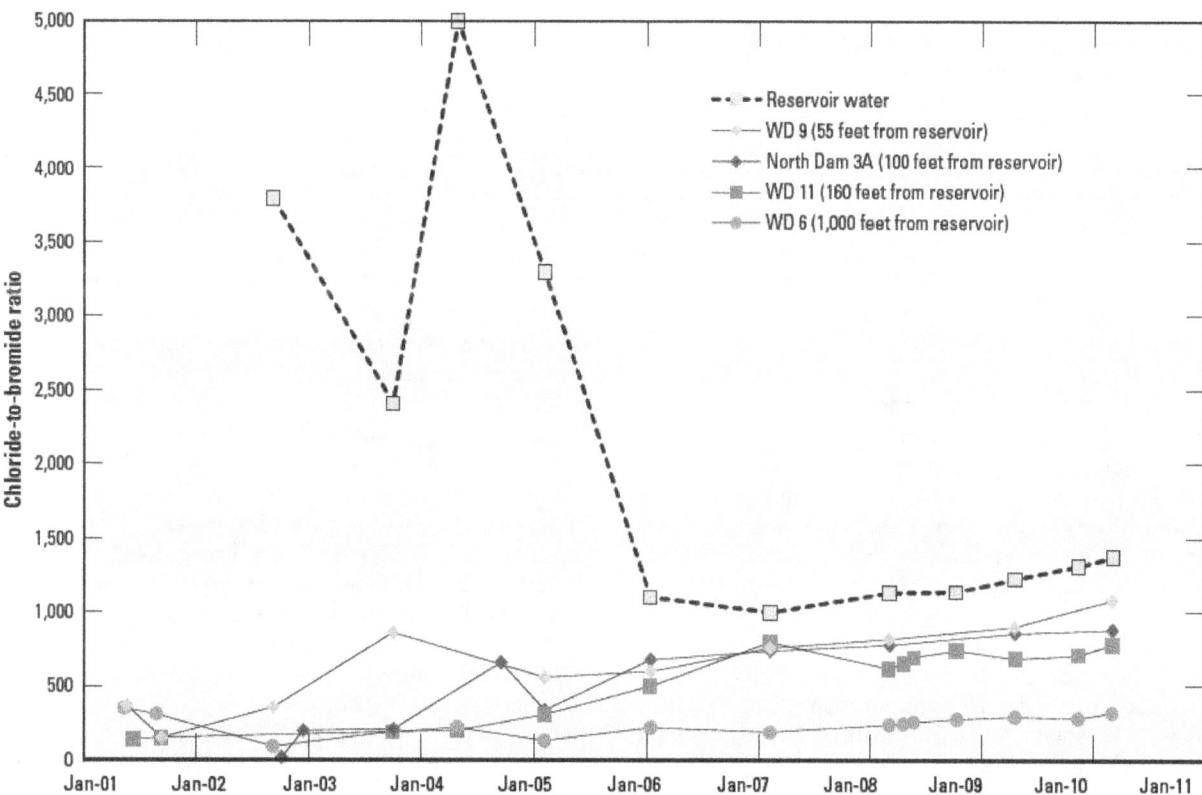

Figure 15.   Chloride-to-bromide ratios of reservoir water and groundwater from selected monitoring wells in Sand Hollow, Utah.

Groundwater arsenic concentrations also are being monitored at Sand Hollow. It is possible, theoretically, that recharge of reservoir water having a slightly higher pH (generally around 8.5) compared to native groundwater pH (generally between 7.5 and 8.0), can result in increased dissolution of arsenic occurring naturally in the Navajo Sandstone. Biological processes occurring at the bottom of the reservoir may also increase dissolved organic carbon concentrations and cause reduced conditions that could dissolve and mobilize arsenic. From 2008 to 2010, reservoir water arsenic concentrations ranged from 1.3 to 1.6 µg/L, and pore-water concentrations sampled from temporary piezometers installed in sediments beneath the reservoir ranged from 1.7 to 15.2 µg/L. Arsenic concentrations in groundwater unaffected by reservoir recharge ranged from 2.0 to 28.9 µg/L, whereas concentrations in groundwater affected (or possibly affected) by reservoir recharge were generally higher, ranging from 3.0 to 90.1 µg/L (table 4). Groundwater arsenic concentrations during 2008 and 2009 generally have remained stable and similar to previously reported values (table 1 of Heilweil and others, 2009). Because of the large amount of variability in natural groundwater arsenic concentrations, however, arsenic is not currently being used as a tracer of reservoir recharge.

Dissolved Organic Carbon, Isotopes, and Dissolved Gases

Dissolved organic carbon (DOC) is another potential tracer of reservoir recharge that has been included in Sand Hollow water-quality sampling since 2008. Measured values of DOC in Sand Hollow reservoir water ranged from 2.9 to 5.7 mg/L (table 3). DOC increases as recharge from the reservoir passes through an organic-rich layer that has formed at the sediment-water interface above the pre-existing soils and sandstone in the basin. DOC measured in pore waters collected from temporary piezometers installed in sediments beneath the reservoir ranged from 4.0 to 13.5 mg/L (table 3). Similarly, sediment cores collected from beneath the reservoir during the summer of 2009 had total organic carbon concentrations ranging from 0.9 to 6.4 grams per kilogram (g/kg) and extracted pore-water DOC concentrations from 22 to 130 mg/L (these cores were retrieved in clear 7.0-cm diameter by 61-cm long polycarbonate core barrels using a slide hammer percussion coring device). In contrast, native groundwater at monitoring wells not yet affected by reservoir recharge (WD 4, WD 5, WD RJ, WD 16, WD 18, WD 20) generally had relatively low DOC concentrations of 0.4 to 0.8 mg/L. Groundwater affected by reservoir recharge (North Dam 3A, WD 6, WD 9, WD 11) had DOC concentrations of 1.2 to 1.9 mg/L, indicating a mix of native groundwater and reservoir recharge.

Tritium concentrations in reservoir water and groundwater have been measured sporadically since 1999. Tritium is a radioactive isotope of hydrogen with a half-life of 12.43 years and can occupy one of the hydrogen sites in a water molecule. It is produced naturally in small amounts as cosmic rays interact with gases in the upper atmosphere, and was produced in larger amounts during the 1960s with above-ground nuclear testing (Solomon and Cook, 2000). Like DOC, surface-water tritium concentrations are generally higher than in groundwater, making it a useful tracer of reservoir recharge. Both reservoir water and pore water collected from shallow piezometers beneath the reservoir had tritium concentrations of 2.5 to 4.6 TU (table 3). In contrast, native groundwater in monitoring wells, sampled either before the reservoir was constructed or those sampled more recently but away from the reservoir, had tritium concentrations generally less than about 0.5 TU (fig. 16), indicating that the majority of natural recharge occurred prior to the 1960s. Monitoring wells sampled near the reservoir that were sampled during 2010 (North Dam 3A, WD 6, WD 9, WD 11) had tritium concentrations of 2.8 to 3.2 TU, indicating the arrival of reservoir recharge. The slightly elevated tritium concentration at WD 15 (0.7 TU) likely indicates the arrival of some natural recharge rather than reservoir recharge. This interpretation is supported by its low Cl:Br ratio (< 200) and its shallow screened depth (38 to 58 ft). This well is located in an area that was previously categorized as a "medium" natural recharge zone, compared to many other monitoring wells located in areas categorized as "low" recharge (Heilweil and McKinney, 2007).

Oxygen-18 (^{18}O) and deuterium (^{2}H) are stable isotopes of the water molecule that also were investigated as potential tracers of managed aquifer recharge. It was hypothesized that these constituents would show an evaporative shift in the reservoir water and this signal could be used to differentiate between reservoir recharge and natural groundwater in the Navajo Sandstone aquifer. Previously reported isotopic data (Heilweil and others, 2009, table 1) and additional samples collected during 2008 do not consistently show an evaporative shift. This can be due to the time of sample collection (less reservoir evaporation occurs during the winter) and/or the relative amount of evaporation compared to stored surface water in the reservoir. Stable isotopes, therefore, currently are not being used as a tracer of recharge from the reservoir.

Chlorofluorocarbons (CFC–11, CFC–12, and CFC–113) are trace atmospheric gases (synthetic halogenated alkanes) developed in the 1930s as safe alternatives to ammonia and sulfur dioxide for refrigeration (Plummer and Busenberg, 2001). Since the 1930s, these gases have been in the atmosphere and have dissolved into rainwater and surface water. Since 2008, CFCs have been included in Sand Hollow water-quality sampling efforts as additional tracers of reservoir recharge. CFC–12 is considered the most stable of the three chlorofluorocarbons; both CFC–11 and CFC–113 are more likely to be degraded by microbes. Dissolved CFC–12 concentrations in surface water collected from Sand Hollow Reservoir gradually increased from about 1.5 to 3.3 pmol/kg between 2008 and 2010 (table 3). This could have been caused by a relative increase in the ratio of precipitation runoff to groundwater discharge in the Virgin River above the diversion to Sand Hollow Reservoir, associated with above-average precipitation from late 2007 through early 2009. In contrast, older native groundwater in monitoring wells (not receiving post-1950s natural recharge), sampled either before the reservoir was constructed or more recently from wells located farther away from the reservoir, generally had CFC–12 concentrations less than about 0.6 pmol/kg (fig. 17), indicating that the vast majority of natural recharge occurred prior to the 1960s. An exception is WD 15 (discussed above), with CFC–12 concentrations of about 1.2 pmol/kg. Monitoring wells sampled near the reservoir during 2010 (North Dam 3A, WD 6, WD 9, WD 11) had CFC–12 concentrations ranging from 1.6 to 2.9 pmol/kg, signifying the arrival of reservoir recharge.

Sulfur hexafluoride (SF_6) is another atmospheric gas tracer, which, since the 1970s, primarily comes from the industrial production of high-voltage electrical switches (http://water.usgs.gov/lab/sf6/background/). SF_6 was sampled during 2008 and 2009 at Sand Hollow as a tracer of reservoir recharge. Measured SF_6 concentrations in surface water from the reservoir, representing water equilibrated with modern atmospheric concentrations of the gas, ranged from about 1.5 to 2.0 fmol/kg (table 3). Groundwater SF_6 concentrations ranged from about 0.1 to 3.5 fmol/kg. While patterns of low versus high groundwater SF_6 concentrations were generally similar to CFC–12 concentrations, anomalously high SF_6 concentrations were measured at WD 11 and WD 18. The concentration in 2009 at WD11, 160 ft from the reservoir, was 3.5 fmol/kg. This "excess" SF_6 is consistent with the high TDG pressures and can be explained by dissolution of trapped air beneath the reservoir. A high SF_6 concentration (1.5 fmol/kg) was also measured in 2009 at WD 18, located far from the reservoir. This was similar to concentrations in the reservoir, but it is assumed that this well does not yet contain modern water, based on low concentrations of tritium and chlorofluorocarbons. This "excess" SF_6 is instead attributed to the dissolution of air bubbles that formed within the aquifer because air was used as the drilling fluid; these air bubbles containing modern atmospheric SF_6 should dissipate with time. In contrast, at other wells with older native groundwater, background SF_6 concentrations were significantly lower. For example, at WD 4 and WD 5 (older monitoring wells constructed in 1995) concentrations ranged from 0.1 to 0.4 fmol/kg (table 3). SF_6, therefore, generally could be useful as a tracer of reservoir recharge at Sand Hollow.

Water samples collected at selected monitoring wells also were analyzed for the dissolved noble gases helium, neon, argon, krypton, and xenon. Initial samples were collected prior to completion of Sand Hollow Reservoir; additional samples were collected sporadically between 2002 and 2009. Similar to TDG pressure, neon excess is a good indicator of trapped air in the Navajo Sandstone aquifer. While all of the noble-gas concentrations can be useful for evaluating recharge processes, neon excess is the simplest metric to calculate and the most straight-forward to interpret. Neon excess refers to the amount of dissolved neon in water above atmospherically equilibrated

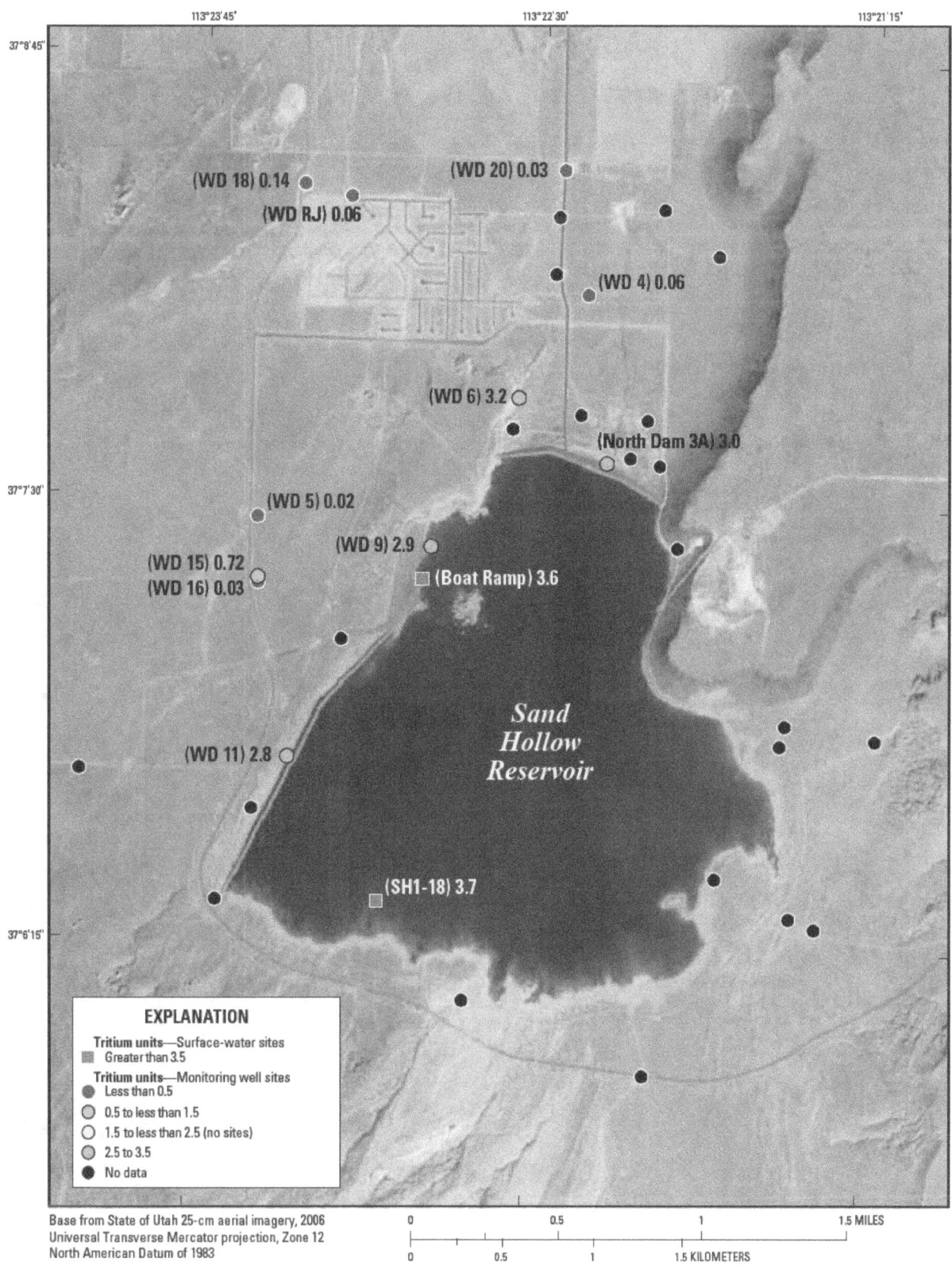

Figure 16. Tritium concentrations in reservoir water and groundwater from selected monitoring wells in Sand Hollow, Utah, March 2010.

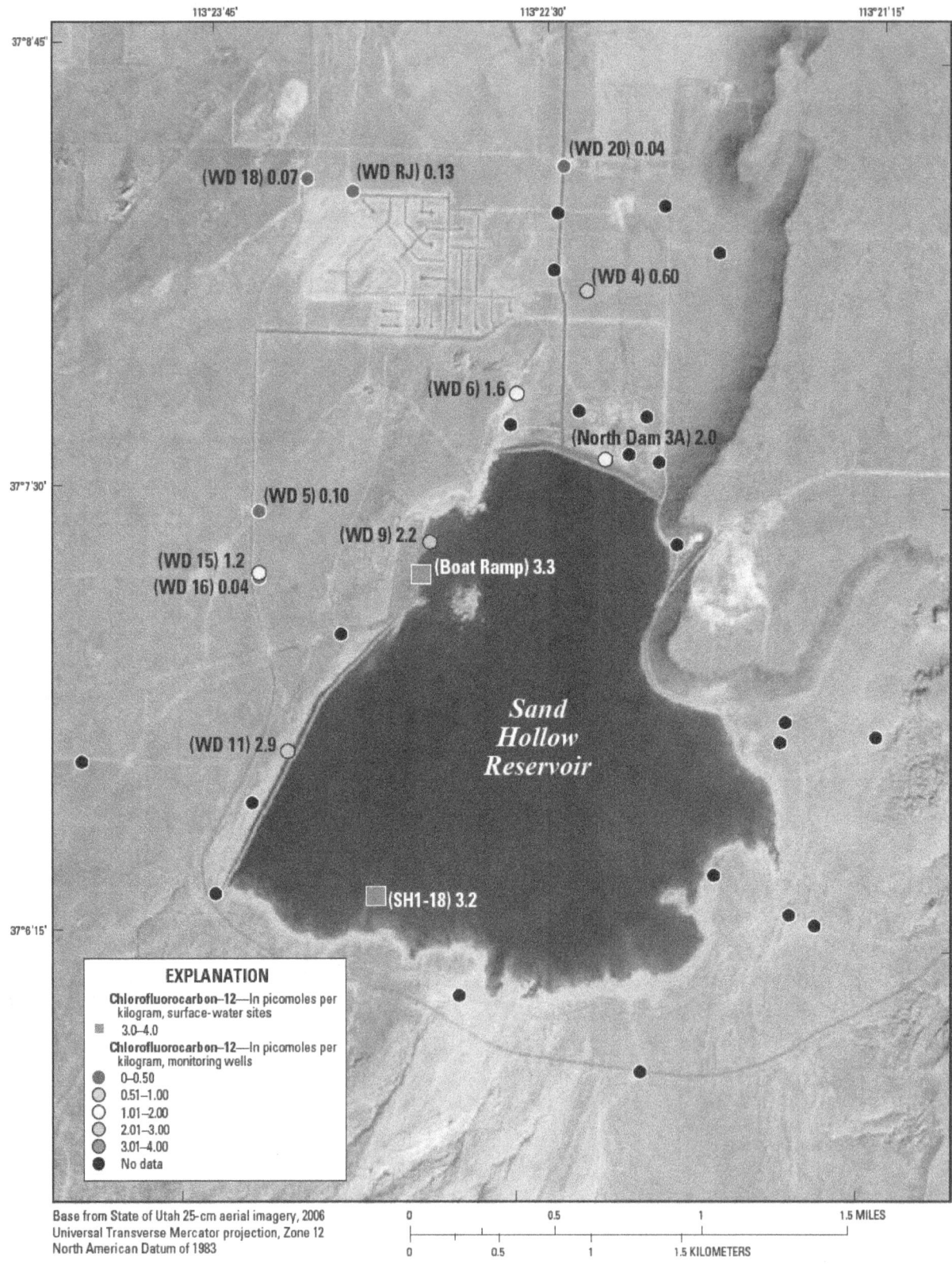

Figure 17.   Chlorofluorocarbon–12 concentrations in reservoir water and groundwater from selected monitoring wells in Sand Hollow, Utah, March 2010.

amounts. High values of neon excess indicate that, as groundwater levels rose during the initial filling of the reservoir, air bubbles were present in the sediments and underlying sandstone. Groundwater that is under high hydrostatic pressure and passes by these trapped air bubbles, dissolves neon and the other noble gases. Prior to the filling of Sand Hollow Reservoir, noble-gas measurements in monitoring wells indicated excess neon concentrations from –0.2 to 71 percent (table 3). In contrast, peak excess neon concentrations, indicating the arrival of recharge from the reservoir, have reached values ranging from about 160 to 320 percent in monitoring wells located in proximity to the reservoir. Thus, neon excess also is a useful tracer of recharge from Sand Hollow Reservoir.

Evaluating the Arrival of Managed Aquifer Recharge at Monitoring Wells

Changes in values, or more specifically, the arrival of peak values for selected field parameters (TDG pressure, DO, and specific conductance), dissolved chemical constituents (Cl:Br ratios, major-ion chemistry), isotopes (tritium), and dissolved gases (CFC–12, neon) at the monitoring wells were used to evaluate the movement of managed aquifer recharge through the Navajo Sandstone aquifer in Sand Hollow. While the various tracers showed recharge from Sand Hollow Reservoir has arrived at monitoring wells closest to the reservoir, they often indicate different peak arrival years for the same monitoring well (table 5). This is likely due to different behavioral characteristics of each of the tracers, such as adsorption and retardation, dispersion, and gas dissolution as recharge enters

and moves through the aquifer. In general, these tracers indicate that recharge from the reservoir arrived earliest at the wells located closest to the reservoir (WD 9, North Dam 3A). Although North Dam 3A is located a little farther from the reservoir than WD 9, it is a much shallower well (initially a dry borehole in the vadose zone) and received recharge from the reservoir in mid-2002, when it first became saturated (even though the peak Cl:Br ratio and DO were not measured until 2004 and 2005, respectively). The recharge arrival year at WD 9 is less certain, with tracer peaks occurring between 2003 and 2006. At WD 11, all of the tracers, except neon excess, indicated that the peak breakthrough of recharge from the reservoir likely occurred in 2005 or 2006. At WD 6, the majority of the tracers indicated that recharge arrived between 2005 and 2009. Although CFC and tritium concentrations were still rising at WD 6, this may partially reflect their rising concentrations in the reservoir in recent years; tritium is not a meaningful tracer of reservoir recharge at this well because of the high natural background concentrations of 4.8 to 6.9 TU measured in 2001 prior to the reservoir. While various tracers sampled at WD 15 (a shallow monitoring well screened from 38 to 58 ft below land surface) are elevated above background levels, this is likely caused by natural recharge and the rising water table rather than indicating the arrival of reservoir recharge. WD 15 is located about 2,400 ft west of the reservoir, much farther than the other monitoring wells showing arrival of reservoir recharge. This interpretation is consistent with tracer concentrations at WD 16 (located at the same site but screened from 282 to 302 ft below land surface), which do not indicate the arrival of reservoir recharge.

Table 5.   Summary of tracer peaks showing year of arrival for reservoir recharge at selected monitoring wells in Sand Hollow, Utah.

[Cl:Br, chloride-to-bromide ratio; —, no data available; ?, uncertain year of tracer peak; NMF, not meaningful]

Site name	Distance from reservoir, in feet	Total dissolved-gas pressure	Dissolved oxygen	Specific conductance	[1]Cl:Br	Major-ion chemistry	Tritium	CFC-12	Neon excess
WD 9	55	2003, 2004-05?	2003, 2004?	2006	Rising	prior to 2009	prior to 2008	prior to 2008	2003-05?
North Dam 3A	100	—	2005	2002	Rising	2002	2002	prior to 2008	—
WD 11	160	2004, 2005-08?	2004-05	2004	Rising?	prior to 2009	prior to 2008	prior to 2009	2008?
WD 6	1,000	2009	2009	2008	Rising	prior to 2009	[3]NMF	Rising	2009
WD 15 (shallow)	2,400	[2]Elevated	[2]Elevated	[2]Elevated	No arrival	No arrival	[2]Elevated	[2]Elevated	[2]Elevated
WD 16 (deep)	2,400	No arrival	No arrival	No arrival	No arrival	No arrival	No arrival	No arrival	No arrival
WD 4	2,600	—	No arrival	No arrival	No arrival	No arrival	No arrival	No arrival	No arrival
WD 5	2,800	—	No arrival	No arrival	No arrival	No arrival	No arrival	No arrival	No arrival
WD 20	5,000	No arrival	No arrival	No arrival	No arrival	No arrival	No arrival	No arrival	No arrival
WD RJ	5,200	No arrival	No arrival	No arrival	No arrival	No arrival	No arrival	No arrival	No arrival
WD 18	5,900	No arrival	No arrival	No arrival	No arrival	No arrival	No arrival	No arrival	No arrival

[1]Based on data from both Table 4 of this report and Table 1 of Heilweil and others, 2009.

[2]Elevated total dissolved-gas pressure, dissolved oxygen, specific conductance, tritium, CFC-12, and neon excess during 2009–2010 indicate local natural recharge and rising water table (dissolution of trapped air bubbles and chloride bulge) rather than reservoir recharge.

[3]Not meaningful tracer of reservoir recharge because of elevated tritium concentrations prior to reservoir construction.

Summary

The objectives of this study were to both quantify amounts of recharge from Sand Hollow Reservoir and to evaluate its movement through the Navajo Sandstone aquifer, updated to conditions in 2010. This study follows previous USGS Scientific Investigations Reports 2005–5185 (Heilweil and others, 2005), 2007–5023 (Heilweil and Susong, 2007), and 2009–5050 (Heilweil and others, 2009).

Since its inception in 2002, diversions to the reservoir from the nearby Virgin River resulted in a generally rising water-level altitude, from about 3,000 ft in 2002 to a maximum of about 3,060 ft in 2006, and then fluctuated between about 3,040 and 3,060 ft during 2008 and 2009. Similarly, ground-water levels in monitoring wells closest to the reservoir generally rose between 2002 and 2006, and then fluctuated with reservoir altitude and nearby pumping from production wells. Water levels in monitoring wells farther from the reservoir were still rising through 2009.

About 13,000 acre-ft of groundwater were withdrawn between 2004 and 2009, mostly from production wells located near the North Dam. French drains, installed to capture shallow seepage near the reservoir, are also pumped as they fill with water. About 4,800 acre-ft of groundwater were pumped from the North Dam drain between 2003 and 2009. This water initially was returned to the reservoir, but since 2007 has been used by Sand Hollow Resort for irrigation. About 800 acre-ft of water were pumped from the West Dam drain back into the reservoir from 2005 through 2009. In 2006, the West Dam Spring drain was constructed and has largely replaced the function of the West Dam drain. About 8,500 acre-ft have been pumped from this drain from 2006 through 2009 into the WCWCD's municipal supply system.

Total annual surface-water inflow to Sand Hollow Reservoir has ranged from about 800 acre-ft in 2007 to 56,000 acre-ft in 2005. Total inflow from 2002 through 2009 was about 154,000 acre-ft. The general increase in reservoir water-level altitude and surface area from 2002 and 2007 resulted in a steady increase in the volume of annual evaporation from about 1,000 to about 6,600 acre-ft through 2006, then leveled off from 2007 through 2009. Total estimated cumulative evaporative loss from 2002 through 2009 was about 37,000 acre-ft. During this same period, annual reservoir recharge to the underlying Navajo Sandstone aquifer fluctuated between about 5,000 and 18,000 acre-ft. In 2009, recharge was approximately 11,000 acre-ft. Total calculated reservoir recharge from 2002 through 2009 was about 86,000 acre-ft with a 2 standard deviation uncertainty of 9,600 acre-ft. From 2002 through 2009, calculated monthly recharge volumes ranged from about 200 to 3,500 acre-ft, and average daily recharge rates (calculated for each month) ranged from 0.001 to 0.43 feet. From March 2002 through May 2002, there was a rapid decrease in rates as the vadose zone wetted up and the reservoir became hydraulically connected to the aquifer. From mid-2002 through mid-2007, there was a gentler decline in recharge rates, likely caused by both the decreasing hydraulic gradient in the aquifer and clogging beneath the reservoir. From mid-2007 through 2009, recharge rates stabilized.

Water-quality sampling was conducted at various monitoring wells in Sand Hollow to evaluate the timing and location of reservoir recharge moving through the aquifer. Tracers of reservoir recharge include major and minor dissolved inorganic ions, tritium (a radioactive isotope of hydrogen), dissolved organic carbon, and dissolved gases, including chlorofluorocarbons, sulfur hexafluoride, and noble gases. The various tracers, however, often have different peak arrival years at individual monitoring wells. This is likely due to different behavioral characteristics of each of the tracers, such as adsorption and retardation, dispersion, and gas dissolution as recharge enters and moves through the aquifer. By 2010, reservoir recharge clearly arrived at monitoring wells within about 1,000 ft of the reservoir. In contrast, these tracers indicate that reservoir recharge has not reached monitoring wells located about 0.5 mi away from the reservoir.

References Cited

Cordova, R.M., 1981, Ground-water conditions in the upper Virgin River and Kanab Creek basins area, Utah, with emphasis on the Navajo Sandstone: State of Utah Department of Natural Resources Technical Publication 70, 87 p.

Heilweil, V.M., Freethey, G.W., Stolp, B.J., Wilkowske, C.D., and Wilberg, D.E., 2000, Geohydrology and numerical simulation of ground-water flow in the central Virgin River basin of Iron and Washington Counties, Utah: Utah Department of Natural Resources Technical Publication 116, 182 p.

Heilweil, V.M., and McKinney, T.S., 2007, Net-infiltration map of the Navajo Sandstone outcrop area in western Washington County, Utah: U.S. Geological Survey Scientific Investigations Map 2988.

Heilweil, V.M., McKinney, T.S., Zhdanov M.S., and Watt, D.E., 2007, Controls on the variability of net infiltration to desert sandstone: Water Resources Research, v. 43, W07431. DOI:10.1029/2006WR005113, 15 p.

Heilweil, V.M., Ortiz, G., and Susong, D.D., 2009, Assessment of managed aquifer recharge at Sand Hollow Reservoir, Washington County, Utah, updated to conditions through 2007: U.S. Geological Survey Scientific Investigations Report 2009–5050, 19 p.

Heilweil, V.M. and Solomon, D.K., 2004, Millimeter–to kilometer–scale variations in vadose-zone bedrock solutes: implications for estimating recharge in arid settings, in Phillips, F., Scanlon, B., and Hogan, J., eds., Ground-water recharge in a desert environment: the southwestern United States, Water Science and Application 9: Washington, D.C., American Geophysical Union, p. 49–67.

Heilweil, V.M., Solomon, D.K., and Gardner, P.M., 2006, Borehole environmental tracers for evaluating net infiltration and recharge through desert bedrock: Vadose Zone Journal, v. 5, p. 98–120.

Heilweil, V.M., Solomon, D.K., Perkins, K.S., and Ellett, K.M., 2004, Gas-partitioning tracer test to quantify trapped gas during recharge: Ground Water, v. 42, no. 4, p. 589–600.

Heilweil, V.M. and Susong, D.D., 2007, Assessment of artificial recharge at Sand Hollow Reservoir, Washington County, Utah, updated to conditions through 2006: U.S. Geological Survey Scientific Investigations Report 2007–5023, 14 p.

Heilweil, V.M., Susong, D.D., Gardner, P.M., and Watt, D.E, 2005, Pre- and post-reservoir ground-water conditions and assessment of artificial recharge at Sand Hollow, Washington County, Utah, 1995–2005: U.S. Geological Survey Scientific Investigations Report 2005–5185, 74 p.

Heilweil, V.M. and Watt, D.E., 2011, Trench infiltration for managed aquifer recharge to permeable bedrock: Hydrological Processes 25, DOI: 10.1002/hyp.7833, p. 141–151.

Hurlow, H.A., 1998, The geology of the central Virgin River basin, southwestern Utah, and its relation to ground-water conditions: State of Utah Water Resources Bulletin 26, 53 p.

McGuinness, J.L. and Bordne, E.F., 1971, A comparison of lysimeter-derived potential evapotranspiration with computed values: U.S. Department of Agriculture Technical Bulletin 1472, Agricultural Research Service, Washington D.C., 71 p.

Plummer, L.N. and Busenberg, E., 2001, Chlorofluorocarbons, in Cook, P.G., and Herczeg, A.L., eds., Environmental tracers in subsurface hydrology, Kluwer Academic Publishers, Boston, p. 441–478.

Rosenberry, D.O., Winter, T.C., Buso, D.C., and Likens, G.E., 2007, Comparison of 15 evaporation methods applied to a small mountain lake in the northeastern USA: Journal of Hydrology, v. 340, p. 149–166.

Solomon, D.K. and Cook, P.G., 2000, ^{3}H and ^{3}He, in Cook, P.G., and Herczeg, A.L., eds., Environmental tracers in subsurface hydrology, Kluwer Academic Publishers, Boston, p. 397–424.

Wilde, F.D. and Radtke, D.B., 1998, National field manual for the collection of water-quality data, Field measurements: U.S. Geological Survey Techniques of Water-Resources Investigations, book 9, chap. A6, 233 p.

www.ingramcontent.com/pod-product-compliance
Lightning Source LLC
Chambersburg PA
CBHW081359170526
45166CB00010B/3142